Asymmetric Synthesis

Volume 1

ANALYTICAL METHODS

Asymmetric Synthesis

Volume 1
ANALYTICAL METHODS

Edited by

James D. Morrison

Department of Chemistry
University of New Hampshire
Durham, New Hampshire

1983

ACADEMIC PRESS, INC.
Harcourt Brace Jovanovich, Publishers

Orlando   San Diego   New York
Austin   Boston   London   Sydney
Tokyo   Toronto

ACADEMIC PRESS, INC.
Orlando, Florida 32887

United Kingdom Edition published by
ACADEMIC PRESS, INC. (LONDON) LTD.
24/28 Oval Road, London NW1 7DX

Library of Congress Cataloging in Publication Data

Main entry under title:

Asymmetric synthesis.

    Includes bibliographical references and index.
    Contents: v. 1. Analytical methods / edited by
James D. Morrison.
    1. Stereochemistry. 2. Chirality. 3. Optical
rotation.   I. Morrison, James D., Date
QD481.A78 1983   541.2'23   83-4620
ISBN 0-12-507701-7

PRINTED IN THE UNITED STATES OF AMERICA

   87 88 89   9 8 7 6 5 4 3 2

Dedication

Each volume of this treatise will be dedicated to an asymmetric synthesis pioneer. This first volume is dedicated to Harry S. Mosher, Professor of Chemistry (Emeritus) at Stanford University.

Harry Mosher was born in Salem, Oregon on 31 August 1915. He received his public school education in Salem, graduating from high school in 1933. He remained in Oregon for an A.B. from Willamette University (1937) and an M.S. in organic chemistry from Oregon State College (1939). He then taught briefly at Willamette (1939–1940) before continuing graduate work at Penn State under the direction of Frank Whitmore, receiving a Ph.D. in organic chemistry in 1942. He stayed on at Penn State as an assistant professor during World War II, directing research on production methods for DDT and on synthetic antimalarials.

Professor Mosher began an illustrious career at Stanford University in 1947. In more than 35 years at Stanford he has made major contributions to the understanding of Grignard reactions, heterocyclic compounds, peroxides, natural products, and organic stereochemistry. He is perhaps best known for his work on asymmetric synthesis (reductions with chiral Grignard reagents and metal hydrides), analytical methods for stereochemical analysis (MTPA, "Mosher's acid"), and animal toxins (salamander and puffer fish toxin and frog toxins). He was one of the first to attempt the rational design of chiral reagents to take advantage of nonbonded interactions in transition states as a means of controlling the sense and efficiency of chirality transfer. Professor Mosher has coauthored more than 150 research papers with over 75 graduate and postdoctoral students.

Contents

7. Nuclear Magnetic Resonance Analysis Using Chiral Derivatives
Shozo Yamaguchi

8. Nuclear Magnetic Resonance Analysis Using Chiral Solvating Agents
Gary R. Weisman

9. Nuclear Magnetic Resonance Analysis Using Chiral Shift Reagents
Robert R. Fraser

Contributors

Numbers in parentheses indicate the pages on which the authors' contributions begin.

KENNETH K. ANDERSEN (45), Department of Chemistry, University of New Hampshire, Durham, New Hampshire 03824

JOHN FINN (87), Roger Adams Laboratory, School of Chemical Sciences, University of Illinois, Urbana, Illinois 61801

ROBERT R. FRASER (173), Ottawa–Carleton Institute for Graduate Research in Chemistry, University of Ottawa, Ottawa, Ontario K1N 9B4, Canada

DIANA M. GASH (45), Department of Chemistry, University of New Hampshire, Durham, New Hampshire 03824

JEAN-PAUL GUETTÉ (29), Laboratoire de Chimie Organique, Conservatoire National des Arts et Métiers, 75141 Paris 03, France

GLORIA G. LYLE (13), Division of Earth and Physical Sciences, College of Sciences and Mathematics, The University of Texas at San Antonio, San Antonio, Texas 78285

ROBERT E. LYLE (13), Division of Chemistry and Chemical Engineering, The Southwest Research Institute, San Antonio, Texas 78284

JAMES D. MORRISON (1), Department of Chemistry, University of New Hampshire, Durham, New Hampshire 03824

WILLIAM H. PIRKLE (87), Roger Adams Laboratory, School of Chemical Sciences, University of Illinois, Urbana, Illinois 61801

JOHN D. ROBERTSON (45), Department of Chemistry, University of New Hampshire, Durham, New Hampshire 03824

ALAIN R. SCHOOFS (29), Centre de Recherche Delalande, 92500 Rueil-Malmaison, France

VOLKER SCHURIG (59), Institut für Organische Chemie, Universität Tübingen, D-7400 Tübingen, Federal Republic of Germany

GARY R. WEISMAN (153), Department of Chemistry, University of New Hampshire, Durham, New Hampshire 03824

SHOZO YAMAGUCHI (125), Department of Chemistry, College of General Education, Tohoku University, Sendai 980, Japan

Preface

At one time, asymmetric synthesis was considered to be a rather exotic academic specialty. This is no longer the case. Today it is a primary focus of activity for many of the leading researchers in both the academic and industrial communities. In the years since the first definitive monograph on asymmetric synthesis was published (*Asymmetric Organic Reactions,* 1971, by J. D. Morrison and H. S. Mosher), there has been a literature explosion in the field. It is now so rich and diversified that no single author (or pair of authors) can really review it properly. For that reason an edited multivolume treatise summarizing progress during the period since 1971 was conceived. Thus, in addition to the present volume, we have concurrently in press or in preparation volumes covering stereodifferentiating addition reactions, catalytic processes, and the chiral carbon and heteroatom pool.

This first volume (Analytical Methods) begins with a chapter that summarizes and illustrates the methods used to obtain chiral compounds. Its purpose is to place asymmetric synthesis in the context of other techniques. The following eight chapters show how one can evaluate the chiral efficiency of an asymmetric synthesis or any of the other processes that produce chiral compounds. The analytical methods used by workers in the field are described, in most cases by authors who pioneered their development. These techniques have played a major role in the development of new or more efficient asymmetric syntheses. Whereas only two decades ago the polarimeter was virtually the only tool used to evaluate enantiomeric composition, various methods are now available. Many of the newer methods are fundamentally superior to polarimetry for the accurate determination of enantiomer ratios.

The survey of analytical methods begins with a chapter on polarimetry, which includes a guide to the instruments available commercially. A chapter on competitive reaction methods reviews the techniques developed mainly by Horeau

and co-workers. A chapter on isotope dilution presents this rather old and sometimes confusing technique in a fresh and lucid way, using graphic illustrations.
There are two chapters on exciting new developments in gas and liquid chromatography and three chapters on nuclear magnetic resonance methods (chiral
derivatives, chiral solvating agents, and chiral shift reagents).

This volume is essential reading for chemists in all areas. The concepts and
methods described transcend traditional boundaries of specialization. There
should be special interest on the part of pharmaceutical chemists and others
concerned with the synthesis and analysis of complex chiral molecules.

1

A Summary of Ways to Obtain Optically Active Compounds

James D. Morrison
Department of Chemistry
University of New Hampshire
Durham, New Hampshire

I. Introduction

In this very brief introductory chapter, asymmetric synthesis is placed in perspective with other ways of obtaining chiral compounds. References and examples have been selected to show how each method works, and no attempt has been made to be exhaustive. Almost any of the methods cited could be the subject of a substantial treatise (and some have been).

1

II. Obtaining Chiral Compounds by the FRN Method

This method is the one of choice for most contemporary chemists. I call it the Federal Reserve note (FRN) method, which means that one finds the compound one wants in some specialty chemical catalog and purchases it. Many companies now offer a fairly good selection of chiral compounds. The major problem is that the price is often quite high, but this must be balanced against the fact that many experimental methods for obtaining chiral compounds can be quite time intensive. If student labor (sometimes called slave labor) is available, this may not be a serious consideration. One can always rationalize assigning a tedious resolution, for example, as a valuable "learning experience." Also, if a sufficiently large quantity of chiral material is desired, the cost per gram might be reasonable, even if the labor is fairly expensive. Another problem is that most commercially available chiral compounds must be checked for both chemical purity and enantiomeric composition. Suppliers sometimes specify the former but rarely the latter.

III. Isolating Chiral Compounds from Natural Sources

Citronellal is an example of a chiral compound that is abundant in nature, occurring in over 50 essential oils. The R-enantiomer (1) is the one found naturally, and a common form, Java citronella oil, has a specific rotation $[\alpha]_D^{25}$ of about $+12°$ (neat). This represents a 75–80% enantiomeric excess (ee) (Valentine et al., 1976).

1

The isolation of "natural citronellal" from Java citronella oil is accomplished via the bisulfite derivative. As O'Donnell and Sutherland (1966) pointed out, there are numerous examples of synthetic sequences in the literature in which "natural citronellal" has been used. The failure to obtain crystalline intermediates in some of these syntheses might have been due to the presence of both enantiomers in the starting aldehyde. Recrystallization of the semicarbazone of "natural citronellal" results in an increase in the amount of the R-isomer, and this method can be used to obtain material of higher quality (see also Valentine et al., 1976).

The author has found that a surprising number of chemists believe that *all* chiral compounds isolated from nature are 100% one enantiomer. The case of citronellal is just one example among many showing that this is not so. Nature does not always make just one enantiomer and not the other, although there are many instances in which only one enantiomer has been found.

IV. Resolution of Racemic Modifications by the Physical Separation of Intermediate Diastereomers

Resolution techniques, considered by some to be more art than science, have been the means of isolating many chiral compounds. Since the early 1970s, largely through the efforts of Wilen and Jacques and their co-workers (Jacques *et al.*, 1981), the theory and practical aspects of resolution methods have been painstakingly organized, analyzed, expanded, and explained so that rational approaches likely to be successful can be identified and understood.

It is clear that the study of classical crystallization resolution is continuing. Fairly recent attempts have been made to develop quantitative analysis methods for predicting crystallization resolution behavior (Fogassy *et al.*, 1980). Even more recently, there has appeared a paper describing extensions of the original Pasteur-type manual-separation resolution idea (Addadi *et al.*, 1981).

Many recent studies have been directed toward chromatographic resolution methods. Achiral chromatographic methods can, of course, be used to separate diasteromers and thus effect resolution just as crystallization does. For example, in an interesting application, racemic aldehydic arene–tricarbonylchromium complexes as diastereomeric derivatives prepared from a chiral semicarbazide analog have been resolved chromatographically (Solladie-Cavallo *et al.*, 1979). However, a great deal of effort since the early 1970s has been devoted to chromatographic methods that utilize chiral stationary phases. Chapters 5 and 6 of this volume describe such techniques.

V. Asymmetric Transformations

Asymmetric transformations are often qualified as being either first or second order. Both types involve setting up an equilibrium between enantiomers or epimers. In a first-order transformation of epimers, conditions are found to produce a thermodynamically controlled formation of one of the epimers in excess. In a second-order transformation, one epimer precipitates, driving the equilibrium to produce more of it. The less soluble epimer could be either the more or the less thermodynamically stable one.

A second-order process is, in principle, a very attractive option for obtaining certain chiral compounds. For example, one might start with a 50 : 50 mixture of epimeric salts from a racemic organic acid and an optically active base exactly as in a classical resolution. The objective would be to find conditions that would put these salts in dynamic equilibrium via a configurational change at the acid chiral center and to do this in a solvent system in which one epimer would be less soluble than the other. Under ideal conditions, then, it might be possible to "transform" all of the equilibrating epimer mixture to the less soluble epimer of the salt. Treatment with a stronger acid would yield only one enantiomer of the organic acid, and an overall total conversion of a racemate to a single enantiomer could, in principle, be accomplished. Although total "transformations" are rare, a number of second-order systems produce more than the 50% yield of pure enantiomer that would be possible theoretically (but is never obtained in practice) from a resolution.

Similar principles apply to the "transformation" of an equilibrium between enantiomers under racemizing conditions, in which the equilibrium constant is unity. If such a system reaches supersaturation, it can be seeded with crystals of the desired enantiomer to promote crystallization. The equilibrium then readjusts to provide a 50 : 50 ratio again in the solution or melt.

Seeding is essential in order to gain control of the process. In a study that must rank as one of the most outstanding examples of scholarly patience and fortitude in the chemical literature, Soret investigated the *spontaneous* deposition of $NaClO_4$ from aqueous solution in enantiomeric crystalline forms (Soret, 1901; see also Kipping and Pope, 1898). Solutions left open to the air gave crystals that were mainly left-handed, but in 844 trials sealed samples deposited right-handed crystals 51.3% of the time and left-handed crystals 48.7% of the time.

A true spontaneous deposition of chiral crystals or molecules, like the flipping of a coin, should give the right-handed form as often as the left-handed form, if the number of trials is sufficiently large. When solutions are left open to the air, however, accidental seeding is almost certain to occur, as has been demonstrated for several systems (Pincock and Wilson, 1973). By crystallization of 1,1'-binaphthyl from a melt, highly active compounds can be obtained (Pincock *et al.*, 1971; Pincock and Wilson, 1973, 1975). In 200 crystallizations an excess of R-enantiomer was obtained in 52.5% of the trials. The sum of all the observed specific rotations was $+0.14°$, and the maximum observed rotations were $+206°$ and $-218°$.

A well-studied second-order asymmetric transformation via epimers is the high-yield conversion of racemic methionine or phenylglycine (or substituted phenylglycine) esters to one enantiomer. For example, stirring methyl DL-phenylglycinate with 1 equivalent each of benzaldehyde and D-tartaric acid gave an 85% yield of methyl D-phenylglycinate hydrogen D-tartrate. The benzaldehyde is essential for forming the stereochemically labile imine intermediate. Filtrates from such transformations were reused to raise the overall yield of the D-salt to 95% (Clark *et al.*, 1976).

Trost and Hammen (1973) have pointed out another intriguing system in which the principles of asymmetric transformation could be exploited. Malates or dibenzoyl tartrates of sulfonium cations (e.g., structure **2**) (Trost and Hammen, 1973) that have three different groups attached to the sulfur are stereochemically labile at the sulfur center. The resolution of such salts under epimerizing conditions that cause one epimer to precipitate would constitute a second-order asymmetric transformation.

2 (+)-**Tartrate**

The second-order asymmetric transformation of racemic α-amino-ε-caprolactam (ACL, **3**) to the L-isomer (an L-lysine precursor) is a particularly elegant example (Boyle *et al.*, 1979). The key features of the process are the alkoxide-catalyzed racemization of **3** in the form of nickel chloride complexes [Ni(**3**)$_3$Cl$_2$] and an enantioselective precipitation of Ni(L-**3**)$_3$Cl$_2$ driven by chiral seed crystals. The latter condition makes this process different from the more conventional examples, involving epimers, that were described above.

3

Taking the three chiral bidentate ACL ligands and the chiral Ni atom into account gives 24 possible stereoisomers for the labile Ni(**3**)$_3$Cl$_2$ species. In solution, the desired complexes (homochiral with respect to ACL but not Ni), that is, those with composition Ni(L-**3**)$_3$CL$_2$ (four diastereomers), are present along with an equivalent amount of each Ni(D-**3**)$_3$Cl$_2$ diastereomer. There is also a set of eight diastereomers (each a pair of enantiomers) in which two ACL ligands are enantiomers of the third ACL ligand (heterochiral with respect to ACL). The various species are interconvertible by ligand exchange.

Seed crystals of Ni(L-**3**)$_3$Cl$_2$ cause a selective precipitation of this form from a supersaturated solution. In the alcohol–KOH solution, racemization of the ACL–Ni complexes adjusts the equilibrium to compensate for the loss of the L-ACL. The crystals, mainly an Ni(L-**3**)$_3$Cl$_2$ solvate, are not racemized. They are separated, converted to L-**3**·HCl (with some further enrichment in L over D due to separation of pure enantiomer from racemate), and then converted to L-lysine·HCl by hydrolysis with aqueous hydrochloric acid.

VI. Kinetic Resolution

In the last example of the previous section a kinetic resolution of ACL–Ni complexes was triggered by chiral seeding. In that example, because of racemization, the original 50 : 50 solution ratio of the D- and L-forms of the chiral ligand was regenerated. There are a very large number of examples of kinetic resolutions (biochemical and chemical) in which an enantioselective reaction takes place, but racemic conditions are not reestablished afterward. The efficiency of such a process depends on the relative rates of reaction of the two enantiomers with a chiral reactant, and the maximum yield of one isomer (as in a classical resolution) cannot be greater than 50%.

Mori and Akao (1980) studied the enantioselective hydrolysis of acetates of racemic alcohols in a nutrient broth containing *Bacillus subtilis* var. *niger*. The bacterium preferentially hydrolyzed (S)-acetates to the corresponding alcohols, leaving (R)-acetates unchanged. The degree of enantioselectivity varied considerably depending on the structure of the alcohol. Several alkylalkynylcarbinols and their corresponding acetates were obtained in about 20 to 40% yield with % ee values typically in the 30–90% range.

In a demonstration of how one might use such a method as a first step in the complete resolution of a crystalline alcohol, acetates of racemic ethyl mandelate were kinetically resolved. The bacterial hydrolysis products [(R)-acetoxy ester and (S)-hydroxy ester] were separated and hydrolyzed to (R)- (73% ee) and (S)-mandelic acid (47% ee). Repeated crystallization caused fractionation of the pure enantiomers from the racemate in the enriched samples, and ultimately both pure enantiomers of mandelic acid were obtained (but in very low yields).

Similarly, purified enzymes have been used as reagents for kinetic resolutions. A notable example is α-chymotrypsin, which is effective for the enantioselective hydrolysis of various kinds of esters (Jones and Marr, 1973). Several features of such resolutions, some of which may limit their utility, should be recognized. The enzyme is used in an aqueous solution, but sometimes small amounts of organic solvents are added to improve the solubility of substrates. Resolutions can sometimes be effective, however, even if the hydrolysis reaction mixture is heterogeneous. The pH must be maintained at about 7.8 to 8.0. The scale described for enzymatic kinetic resolutions in the chemical research literature is usually quite small (a few grams). There are, of course, some very large scale commercial processes that employ enzymes as chiral reagents. Fermentations are a well-known example. Immobilized enymes have been employed in large-scale asymmetric processes (Chibata, 1978).

In some cases one can "engineer" the enantioselectivity of enzymes toward multifunctional substrates by using derivatives of a group (or groups) in the substrate that are rather remote from the site of reaction. For example, Tabushi *et al.* (1975) studied the α-chymotrypsin-catalyzed hydrolysis of $PhCH_2CH(OH)COOMe$ and a variety of derivatives acylated at the α-OH group. The k_L/k_D hydrolysis rate ratio varied from 3.6 (*t*BuCO derivative) to 26 (CH_3COOCH_2CO derivative), with unacylated substrate falling at 16.

Some of the most efficient examples of nonbiochemical kinetic resolutions are found in work based on the chiral epoxidation systems developed by Sharpless and co-workers (Martin *et al.*, 1981; Sharpless *et al.*, 1982; Roush and Brown, 1982). These studies are summarized in Volume 4 of this treatise (see also Chapter 3, this volume).

VII. Asymmetric Synthesis

The basic principles and definitions of asymmetric synthesis have been described and illustrated (Morrison and Mosher, 1971 and 1976) and are generally well understood by chemist and biochemists. In a typical asymmetric synthesis, prochiral groupings are converted to chiral groupings. The chapters in this multivolume treatise provide a rich variety of examples of this kind of reaction.

Since *Asymmetric Organic Reactions* was written (Morrison and Mosher, 1971 and 1976), the number of publications in this area has increased dramatically. Even more impressive and important has been the general improvement of our ability to design reactions that give high % ee. The power of analytical methods for evaluating results is also much greater now. In fact, improved analytical ability has been responsible, in part, for advances in the scope and quality of asymmetric synthesis. Freedom from dependence on the polarimeter has made it possible to evaluate new products accurately, quickly, and absolutely. Reference to the early literature for "maximum" rotations and configurational assignments is no longer essential. Now, as part of the almost routine stereochemical analysis of new products, % ee and sometimes the configuration can be obtained, often from the same analytical experiment. Chapters 5–9 in this volume describe most of the useful techniques.

Practical asymmetric syntheses are of two general kinds. The majority involve addition to some unsaturated grouping and the selection of enantiotopic or di-astereotopic faces. Less commonly, there is a selective substitution or modification of paired ligands at a prochiral center. As specific examples of these common strategies, consider the following two cases in which both types of selection occur in the overall asymmetric process.

In an intriguing paper that often has been overlooked in reviews of asymmetric synthesis, Job and Bruice (1974) reported not only some experiments rich in chiral chemistry, but also one of the first chiral "breeder reactions" (although it was not advertised as such). By the use of chiral tetraamine ligands **4** and **5**

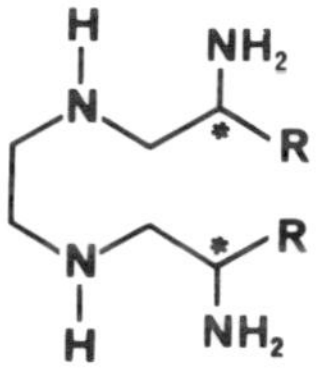

4 R=Me

5 R=_i_Pr

prepared from (S)-alanine and (S)-valine, respectively, chiral cobalt complexes (Λ-diastereomers) were prepared. These complexes, (**6**) were found to bind aminomethylmalonic acid through the amino group and the (*pro-R*)-carboxylate so as to produce an R-configuration at the α-carbon of the aminomalonate ligand. Decarboxylation in acidic solution gave a cobalt–tetraamine complex of alanine (**7**); the alanine product was 65% S and 35% R when the (S)-alanine-derived tetraamine ligand (**4**) was used and about 68% S and 32% R for the (S)-valine-derived ligand (**5**). Thus, the overall process is a breeder reaction that produces a 30% ee of (S)-alanine using a cobalt(III) complex that incorporates a chiral ligand prepared from pure (S)-alanine.

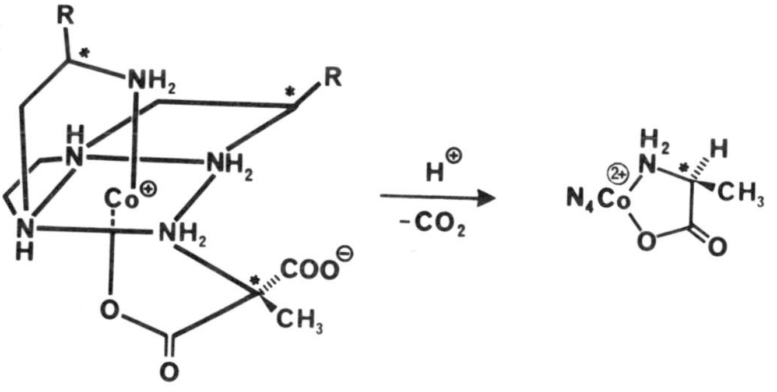

6 (R)   **7 (35R/65S)**

The first step of the process involves the recognition of prochiral carboxylate groups of the aminomethylmalonic acid by the chiral cobalt–tetraamine complex. This recognition appears to be totally selective (CD and X-ray evidence) for binding through the (*pro-R*)-carboxylate (as in **6**). This is a good example of selective modification of paired ligands at a prochiral center. The second step, decarboxylation, probably takes place via an enol-like intermediate (**9**), which is protonated selectively on one of its diastereotopic faces in preference to the other to give a 65/35 ratio of (*S*)- and (*R*)-alanine.

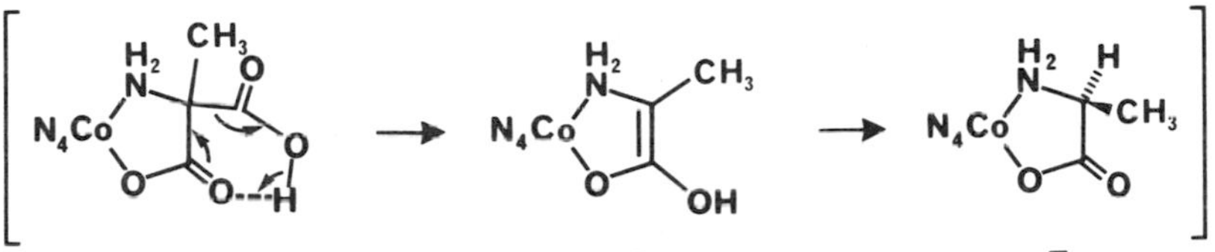

      8   **9**   **7**

Another asymmetric synthesis in which enantiotopic face selection and diastereotopic paired ligand selection are involved simultaneously is the reduction of unsymmetric ketones with chiral Grignard reagents such as **10** (Morrison *et al.*, 1968; Morrison and Ridgway, 1969a). It has been shown that with isopropyl phenyl ketone (and related phenones) there is preferential hydrogen transfer to the *pro-S*-face of the ketone when the Grignard reagent has the *R*-configuration. It has also been demonstrated by deuterium labeling that with the *R*-Grignard reagent there is preferential transfer of the (*pro-S*)-hydrogen.

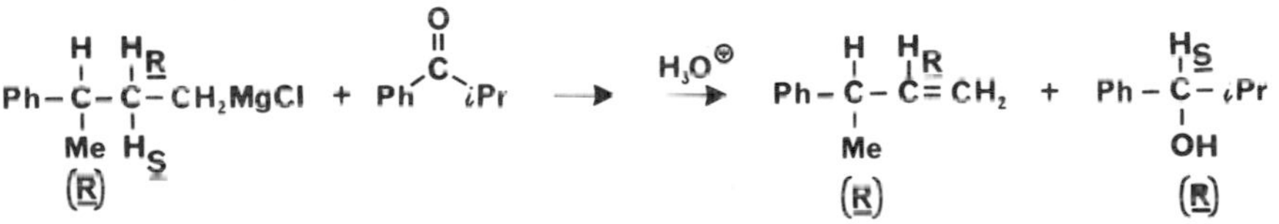

In 1974 Eliel established three guidelines for a good asymmetric synthesis:

1. It must lead to the desired enantiomer in high optical as well as chemical yield.
2. The chiral product must be readily separable from the chiral auxiliary reagent that is needed in the synthesis.
3. Unless the chiral reagent is very much less expensive than the desired product, it must be possible to recover the auxiliary reagent in good yield and in undiminished optical purity.

We might add to the last point the thought that, if the chiral reagent is a catalyst capable of functioning efficiently at a low catalyst/substrate ratio (i.e., the turnover is high), recovery of the reagent might not be essential.

One should keep Eliel's criteria in mind when contemplating an asymmetric synthesis, but their application will, of course, require judgment. Phrases such as high optical and chemical yield, readily separable, very much less expensive, and good yield beg for precise definitions. What is a high optical yield? Only about 20 years ago, 50% ee was rather exciting. Today, greater than 90% ee is becoming commonplace. Will we draw the line at 95% ee in a few years? Eliel's guidelines have an enduring quality because they do not specify absolute levels of performance. As our expectations for asymmetric synthesis become greater we can simply make our quantitative goals more demanding.

While striving for 100% ee one should remember that some special precautions are necessary. Enantiomer ratios can fall prey to subtle influences that might mask the actual efficacy of an asymmetric synthesis. When diastereomers intervene in asymmetric syntheses one must be especially alert. For example, the addition of achiral methyl Grignard reagent to benzaldehyde in a chiral ether such as $(+)$-2,3-dimethoxybutane [$(+)$-DMB] might appear to be a clear-cut asymmetric synthesis. The Grignard reagent, as a chiral etherate, should have the capacity to recognize the enantiotopic faces of benzaldehyde. However, there is

$$\text{MeMgI} + \text{PhCHO} \quad \xrightarrow[\text{2. } H_3O^{\oplus}]{\text{1. }(+)\text{-DMB}} \quad \overset{\text{OH}}{\underset{*}{\text{PhCHMe}}}$$

a postaddition phenomenon that distorts this simple interpretation (Morrison and Ridgway, 1969b). The initial product of the addition is a pair of enantiomeric magnesium alkoxides, which, in the chiral medium, can equilibrate via a Meerwein–Ponndorf–Verley mechanism to give disparate amounts of two diastereomeric alkoxide etherates (the possibility of additional chirality at magnesium is ignored). This is an important factor in the overall process, as was demonstrated by simply allowing a racemic alkoxide to equilibrate in $(+)$-DMB.

$$\underset{(\underline{S})}{\overset{\overset{\text{H}}{|} \quad \overset{\text{I}}{|}}{\underset{\overset{|}{\text{Me}}}{\text{Ph}-\text{C}-\text{O}-\text{Mg}}}\bullet (+)\text{-DMB} \quad \underset{K \neq 1}{\overset{\text{PhCHO}}{\rightleftarrows}} \quad \underset{(\underline{R})}{\overset{\overset{\text{Me}}{|} \quad \overset{\text{I}}{|}}{\underset{\overset{|}{\text{H}}}{\text{Ph}-\text{C}-\text{O}-\text{Mg}}}\bullet (+)\text{-DMB}$$

A more obvious pitfall is the fractionation of stable diastereomeric intermediates by crystallization, washing, or selective reaction. The fractionation of pure enantiomers from racemates by crystallization must also be avoided. For example, a careful study of the reaction sequence (Harada and Okawara, 1973) revealed the importance of such considerations. The intrinsic diastereoselectivity for HCN addition to the chiral imine **10** gave a modest % ee (40–60%) for **13**. The % ee could be boosted to nearly 100% by isolating and washing the diastereomeric acid intermediate (12) and crystallizing the hydrochloride of the final chiral product (13).

$$RCH=NR^* \xrightarrow{\text{HCN}} R\overset{*}{C}HNHR^* \xrightarrow{\text{H}_3\text{O}^{\oplus}} R-\overset{*}{C}HNHR^* \xrightarrow[\text{cat.}]{\text{H}_2} R-\overset{*}{C}H-NH_2$$
$$\qquad\qquad\quad \underset{\text{CN}}{|} \qquad\qquad\quad \underset{\text{COOH}}{|} \qquad\quad \underset{\text{COOH}}{|}$$

$$(\overset{*}{R}-NH_2 = \text{chiral amine})$$

| 10 | 11 | 12 | 13 |

The purpose of being careful to determine the true % ee associated with the enantio- or diastereoselective step of an asymmetric process is to engineer it to a higher level intelligently. For example, the first step of the reaction sequence is subject to manipulation by changing R and R*. If the % ee of 13 is used to probe the effect of such manipulation, however, one should make certain that there is no fractionation along the way.

How asymmetric synthesis will evolve is difficult to predict. The chapters in this multivolume treatise show how far we have come since *Asymmetric Organic Reactions* was written. Progress has been dramatic and will undoubtedly continue. It would appear that we are entering an era of chiral reagents with overwhelming intrinsic prochiral face selectivity. The beauty and power of many of these new reagents (Evans *et al.*, 1982) lie in their capacity to move us beyond the influence of existing chiral centers in substrates to very high, predictable control of stereoselectivity, control that is designed into the chiral reagent.

References

Addadi, L., Gati, E., and Lahav, M. (1981). *J. Am. Chem. Soc.* 103, 1231.

Boyle, W. J., Jr., Sifniades, S., and Van Peppen, J. F. (1979). *J. Org. Chem.* 44, 4841.

Chibata, I. (ed.) (1978). "Immobilized Enzymes, Research and Development." Halstead, New York.

Clark, J. C., Phillips, G. H., and Steer, M. R. (1976). *J. Chem. Soc. Perkin Trans. 1*, p. 475.

Eliel, E. (1974). *Tetrahedron* 30, 1503.

Evans, D. A., Nelson, J. V., and Taber, T. R. (1982). *Top. Stereochem.* 13, 1.

Fogassy, E., Lopata, A., Faigl, F., Darvas, F., Acs, M., and Tokes, L. (1980). *Tetrahedron Lett.* 21, 647.

Harada, K., and Okawara, T. (1973). *J. Org. Chem.* 38, 707.

Jacques, J., Collet, A., and Wilen, S. H. (1981). "Enantiomers, Racemates and Resolutions." Wiley (Interscience), New York.

Job, R. C., and Bruice, T. C. (1974). *J. Am. Chem. Soc.* 96, 809.

Jones, J. B., and Marr, P. W. (1973). *Tetrahedron Lett.*, p. 3165.

Kipping, F. S., and Pope, W. J. (1898). *Trans. Chem. Soc.*, p. 606.

Martin, V. S., Woodard, S. S., Katsuki, T., Yamada, Y., Ikeda, M., and Sharpless, B. (1981). *J. Am. Chem. Soc.* 103, 6237.

Mori, K., and Akao, H. (1980). *Tetrahedron* **36,** 91.

Morrison, J. D., and Mosher, H. S. (1971). "Asymmetric Organic Reactions." Prentice-Hall, Englewood Cliffs, New Jersey; revised ed. (1976), ACS Books, Washington, D.C.

Morrison, J. D., and Ridgway, R. W. (1969a). *J. Am. Chem. Soc.* **91,** 4601.

Morrison, J. D., and Ridgway, R. W. (1969b). *Tetrahedron Lett.,* p. 573.

Morrison, J. D., Black, D. L., and Ridgway, R. W. (1968). *Tetrahedron Lett.,* p. 985.

O'Donnell, G. W., and Sutherland, M. D. (1966). *Aust. J. Chem.* **19,** 525.

Pincock, R. E., and Wilson, K. R. (1973). *J. Chem. Educ.* **50,** 455.

Pincock, R. E., and Wilson, K. R. (1975). *J. Am. Chem. Soc.* **97,** 1474.

Pincock, R. E., Perkins, R. R., Ma, A. S., and Wilson, K. R. (1971). *Science (Washington, D.C.)* **172,** 1018.

Roush, W. R., and Brown, R. J. (1982). *J. Org. Chem.* **47,** 1371.

Sharpless, K. B., Behrens, C. H., Katsuki, T., Lee, A. W. M., Martin, V. S., Takatani, M., Viti, S. M., Walker, F. J., and Woodard, S. S. (1983). *Pure Appl. Chem.* **55,** 589.

Solladie-Cavallo, A., Solladie, G., and Tsamo, E. (1979). *J. Org. Chem.* **44,** 4189.

Soret, C. H. (1901). *Z. Krystallogr. Mineral.* **34,** 630.

Tabushi, I., Yamada, H., and Sato, H. (1975). *Tetrahedron Lett.,* p. 309.

Trost, B. M., and Hammen, R. F. (1973). *J. Am. Chem. Soc.* **95,** 962.

Valentine, D., Jr., Chan, K. K., Scott, C. G., Johnson, K. K., Toth, K., and Saucy, G. (1976). *J. Org. Chem.* **41,** 62.

2

Polarimetry

Gloria G. Lyle
Division of Earth and Physical Sciences
College of Sciences and Mathematics
The University of Texas at San Antonio
San Antonio, Texas

Robert E. Lyle
Division of Chemistry and Chemical Engineering
The Southwest Research Institute
San Antonio, Texas

I. Introduction

The optical rotation of a chemical substance at a single wavelength is an important physical property for the study of a chiral system primarily because it allows one to make reference to the comparable data accumulated in the literature since the early 1800s. In addition to its use in the identification of a particular stereoisomer, the optical rotation at a specific wavelength (polarimetry) or over a range of wavelengths (optical rotatory dispersion) and the difference in absorption of right and left circularly polarized light (circular dichroism) are classical spectral methods for determining the enantiomeric excess or optical purity of a sample.[1] Other methods that are of increasing importance for the determination of the enantiomeric excess and, in some cases, the absolute configuration of an isomer are discussed in detail in subsequent chapters in this volume and can be classified as follows:

1. The optical rotation of a partially resolved compound can be compared with the maximum rotation of the pure enantiomer, which has been determined directly or by calculation. Obtaining a pure enantiomer is frequently difficult except by enzymatic destruction of one enantiomer. Calculation of the maximum rotation of a pure enantiomer can be achieved by the "competitive reaction methods" based on unequal rates of reactions occurring through diastereomeric transition states (Chapter 3, this volume). The use of two reciprocal kinetic resolutions can also lead to the assignment of an absolute configuration (Horeau, 1977).

2. The enantiomeric excess of a partially resolved racemic mixture can be determined directly by the following methods:

    a. Isotope dilution. The preparation of the mixture to be analyzed requires the addition of an isotopically labeled racemic compound to a mixture of enantiomers of the same compound having a known optical rotation but unknown enantiomeric purity. Reisolation of the product yields a sample diluted in its isotopic label as well as in rotation. The enantiomeric excess can be calculated with excellent precision (Chapter 4, this volume).

    b. Analysis by gas or liquid chromatography of the diastereomeric mixture formed from the mixture of enantiomers with a chiral reagent of known optical purity (Chapter 5, this volume). Other separation methods, such as thin-layer chromatography, fractional distillation, countercurrent distribution, or electrophoresis can be used if the separation is efficient.

    c. Chromatographic analysis of the partially resolved mixture of enantiomers using a chiral packing of the column. Gas or liquid chromatography has been the method of choice for quantitative determinations (Chapter 6, this volume).

    d. Nuclear magnetic resonance. As an assay method for a mixture of diastereomers, NMR provides the enantiomeric information without

---

[1]Optical purity describes the ratio of the optical rotation of the mixture of enantiomers to that of the pure enantiomer and is usually the same as the enantiomeric excess, which describes the true composition. In some cases, however, the interaction of the racemate with the enantiomer may cause the rotation to be nonlinear with concentration, requiring another method for the determination of the enantiomeric excess (Horeau, 1969).

requiring separation of the diastereomers (Chapter 7, this volume). The use of a chiral solvent for analyzing a mixture of enantiomers has been successfully developed with NMR analysis (Chapter 8, this volume). Chiral shift reagents provide a diastereomeric environment for an enantiomeric mixture, which permits analysis of the mixture by NMR, but recovery of the substrate requires separation of the chiral reagent (Chapter 9, this volume). These methods have the greatest potential for determining the enantiomeric excess.

e. Differential microcalorimetry. This is the least explored method but one of great potential because of its high precision and lack of need for any chiral reagent. It depends on the phase relationship of the enantiomers, which can be established by a comparison of the properties (heat of fusion, melting points) of the mixture of enantiomers with those of the racemic form (Jacques *et al.*, 1981).

This chapter concentrates on the determination of optical rotatory power by polarimetry and the evaluation of polarimetry as one method for the measurement of enantiomeric composition. Polarimetry is defined as the quantitative investigation of a change in the direction of vibration of linearly polarized (or plane-polarized) light during its passage through an optically anisotropic substance or solution.

II. Instrumentation

The measurement of the optical rotation of chemical substances requires relatively inexpensive equipment, and a variety of instruments are available. The specifications of currently available instruments are based largely on the parameters developed by J. B. Biot (1838). The improved monochromators and light sources make it possible to take measurements at multiple wavelengths, and the increased sensitivity of detection methods permits the use of shorter path lengths of cells. Many instruments, however, have hardly changed for over a century and require visual matching of the field of a half-shade device after the plane-polarized light has traversed the chiral substance. The details of instrumentation have been reviewed by Heller and Curmè (1972).

The component parts of the polarimeter are shown in the schematic diagram of Fig. 1, and modifications that have simplified the collection of data and, in most cases, increased the sensitivity of the measurement are described. The angle of rotation is obtained from the graduated circle attached to the analyzer. Because the sensitivity of the eye to changes in light is greatest when the intensity of light in the room is near that transmitted by the instrument, the accuracy of the visual polarimeters depends on the patience of the observer as well as the quality of the

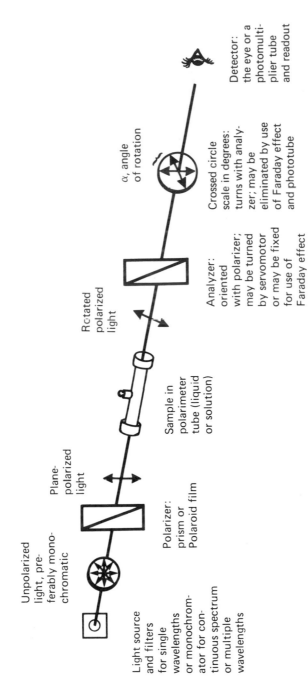

Unpolarized light, preferably monochromatic

Light source and filters for single wavelengths or monochromator for continuous spectrum or multiple wavelengths

Plane-polarized light

Polarizer: prism or Polaroid film

Sample in polarimeter tube (liquid or solution)

Rotated polarized light

α, angle of rotation

Analyzer: oriented with polarizer; may be turned by servomotor or may be fixed for use of Faraday effect

Crossed circle scale in degrees: turns with analyzer; may be eliminated by use of Faraday effect and phototube

Detector: the eye or a photomultiplier tube and readout

Fig. 1. Schematic diagram of a basic polarimeter.

instrument. A 10-min wait in a darkened room is essential before the polarimeter is balanced. A second source of error arises from the polarimeter cells, for the cover plates may be subjected to stress in the tightening of the screw caps, leading to incorrect rotations.

The selection of the instrument for polarimetry depends on the degree of accuracy required by the investigator, the convenience of measurements, and the cost. At least 10 different companies had sales and service available in 1982 in the United States, and comparable or parent companies of some of them are located elsewhere in the world. Other companies, as well as some of these, also specialize in saccharimeters and process instruments for on-line measurement of a flow system. The standard instruments can be grouped as follows:

1. For maximum accuracy, several wavelengths, photomultiplier detection, and digital readout, a group of instruments is available with other options such as chart recorders, printers, or microprocessers.

2. An intermediate-priced group has an accuracy that is approximately as high, but measurements can be made at only a single wavelength and there is only visual detection.

3. A minimal-cost group has the major advantage of easy, fast detection, but there is less accuracy. These instruments can be used for a preliminary measurement before the final readings are made on more accurate instruments.

A summary of the instruments available in the United States is presented in Table I with the characteristics of each model. Some instruments have additional options, and details of the qualifications and cost can be obtained from the respective companies.

A research investigator in the field of asymmetric synthesis must have an instrument of high accuracy but does not necessarily need multiple wavelengths. Photomultiplier detection is convenient and fast and provides a signal that is the input to a microprocessor for data treatment. This feature is particularly convenient for the multiple determinations required by a kinetic study or resolution procedure. The visual polarimeter may suffer from some decrease in sensitivity; however, a patient and careful operator can approach the precision of the electronic detector.

III. Advantages of Optical Rotation Measurements

The principal reason for measuring the optical rotation of a compound is that it provides a comparison with the data for known compounds in the scientific literature and sets a standard reference for new compounds whether isolated from

TABLE I

Specifications of Commercially Available Polarimeters

Company model[a]	Light source, wavelengths	Path length (mm)	Rotation range (deg)	Accuracy (A) and sensitivity (S)	Detector	Readout	Approximate 1981 price (dollars)
Perkin-Elmer 241, 243[b]	Na, Hg lamps, 5 wavelengths	10–100 and microcapacity	±80	A: ±0.002° ± 0.2% > 1°	IP28A photomultiplier	5-digit display with sign	15,000–20,000
JASCO DIP-140, 181[c]	Na, Hg lamps	10–100	±70	A: ±0.002° ± 0.2% > 1°	Photomultiplier	5-digit light-emitting diode for observed or specific rotation	10,000–15,000
Rudolph Autopol II and III[d]	Tungsten–halogen lamp, filters 589 or 546 nm	To 200	±80	A: ±0.002° ± 0.2% > 1°	Photomultiplier	5-digit light-emitting diode	10,000–15,000
Bellingham–Stanley A	Na or Hg lamps, filters	To 40 cm	0–360	A: ±0.01–0.002°	Eye	None	Under 6,000
Bellingham–Stanley P.10	Quartz–halogen, 589.3 nm	To 200	±85	A: ±0.05° S: ±0.02°	Photomultiplier	Digital to 0.1°	Under 5,000
Polyscience SR-6	Na lamp	To 200	±90	A: ±0.1°	Eye	None	About 2,000
Galileo PG-220	Na lamp	25–220	±180	A: ±0.05°	Eye	None	About 3,000
Process and Instruments A[e]	Tungsten lamp filters	1–10 cm	±1.0	S: 0.001°	Photodetector	Digital	Under 2,000
JENA circle polarimeter	Na lamp	100–200	±180	A: ±0.05°	Eye	None	Under 1,000
I²R vertical polarimeter[f]	High-intensity lamp recommended	To 150	±180	A: about ±1°	Eye	None	About 150

[a] For full details of all models, current prices, and options, the reader can contact the following suppliers in the United States: 1, Perkin–Elmer Corporation, Main Avenue, Norwalk, CT 06856. 2, JASCO, Inc., 218 Bay Street, Easton, MD 21601. 3, Rudolph Research, 1 Rudolph Rd., P. O. Box 1000, Flanders, NJ 07836. 4, Epic, Inc., 150 Nassau Street, Suite 1430, New York, NY 10038 (for Bellingham–Stanley models). 5, Polyscience Corporation, 6366 Gross Point Road, Niles, IL 60648. 6, Galileo Corporation of America, 36 Church Street, New Rochelle, NY 10801. 7, Process and Instruments Corporation, 1943 Broadway, Brooklyn, NY 11207. 8, Seiler Instrument and Manufacturing Company, Inc., 170 E. Kirkham Avenue, St. Louis, MO 63119 (for JENA model). 9, Instruments for Research and Industry, Cheltenham, PA 19012 (for I²R model).

[b] Options available for wavelength extension, printout, and kinetic investigations.

[c] Faraday cell modulator, built-in microprocessor for sugar solution readouts, and integration for high-absorbance samples. Options include recorder and digital printer for kinetic investigations and multiple samples.

[d] Options available for digital printout and process control.

[e] Analyzer is moved to give a two-step differential procedure with first reading before unknown and second after unknown.

[f] Useful for rapid approximation measurements and as supplement to the above precision instruments.

natural sources or synthesized. This universal standard is well understood by scientific investigators in biological and medical sciences, geology, and physics as well as by chemists and biochemists. The supplemental use of the newer methods of determining enantiomeric excess is highly desirable, especially for chemical substances of low optical rotatory power. In many cases the low rotation may extend throughout the UV spectrum, and occasionally no rotation may be observed.

Other advantages of this method are the relatively low cost of instrumentation and the ease of measurement. Although the cost of the more expensive polarimeters may exceed $10,000 (in the United States) and measurement throughout the visible–UV region may require a spectropolarimeter of substantially larger cost, the use of NMR spectroscopy for determining enantiomeric excess requires more expensive instrumentation and also chiral adjuvants. Nuclear magnetic resonance is, of course, most versatile in the study of achiral systems and thus does not distinguish between the enantiomers of a racemate unless a chiral environment is provided. Under conditions that allow such distinction between the enantiomers to be made, NMR techniques have advantages over the use of optical rotation for determining the extent of resolution, reaction rates, degree of asymmetric induction, and optical purity. For comparison with data for known compounds or configurational assignments, optical rotation is generally the easiest method to use, at least at present. Unlike NMR, optical rotations are determined on neat liquids or solutions in low-boiling solvents containing only the compound of interest. Recovery of the compound after the determination is therefore easier.

Optical rotation data have one special advantage. With certain classes of natural products the accumulation of data has permitted the formation of structural rules for stereochemical assignment. These rules were initially established by C. S. Hudson (1909, isorotation rules) in the area of carbohydrates, and later rules were developed by Freudenberg et al. (1930, rule of shift), Barton (1945, molecular rotation differences), Mills (1952, 3-substituted cycloalkenes rule), W. Klyne (1954, lactone rule), and J. H. Brewster (1959, atomic asymmetry). For more detailed reviews of monochromatic rotation measurements and rules see Heller and Curmè (1972), Bláha et al. (1971), Schwarz (1964), and Lowry (1935). A useful compilation of data that shows structural relationships of families of compounds and gives optical rotation data and references has been published by Klyne and Buckingham (1978).

IV. Requirements for Precision of Data

The statement that a substance rotates the plane of polarized light always requires supplemental information describing the specific conditions under which it was measured. The standard form for reporting the specific optical rotation of

(R)-(−)-mandelic acid is shown by Elsenbaumer and Mosher (1979) and includes the temperature of measurement, the wavelength (sodium D-line), solvent,

$$[\alpha]_D^{20} - 155.4° \qquad (c = 2.913, \text{ 95\% EtOH; 100\% ee})$$

and the concentration c in grams per 100 ml. The designation of enantiomeric excess (ee) was indicated in this case because it was essential to the synthesis described in the paper and had been established by several methods so that rotation was a reliable reference. A sample can be measured neat, in solution, as a solid film, or as a gas in certain cases. The observed rotation α is a function of the number of molecules through which the plane-polarized light of wavelength λ passes, as in

$$[\alpha]_\lambda^t = \alpha/l\rho, \qquad (1)$$

which expresses specific rotation $[\alpha]_\lambda^t$ for neat liquids, where l is path length in decimeters and ρ the density of the liquid or solid in grams per liter at the temperature of the experiment. The formula for solutions is discussed in Section IV,B. Measurement of the optical rotation is assumed by many chemists to be a trivial experimental procedure because the basic instrument is relatively simple and the process is readily adapted to undergraduate laboratories. The fact is, however, that optical rotations are not necessarily self-consistent because variations can occur with any of the parameters often assumed to be constants. These include temperature, concentration, wavelength, and solvent, which are discussed in the sequel. Plattner and Heusser (1944) suggested that the errors in measurement of optical rotation resulting from temperature and concentration effects were at least ±4%. Although more precise measurement is possible with modern instrumentation, the values are no better today if the temperature and concentration effects are ignored.

A. Variation of Optical Rotation with Temperature

The effect of temperature on the observed rotation results from at least three factors: (a) the volume of the solvent or liquid changes with temperature, thus changing the number of molecules in the optical path; (b) the interaction of the molecules with the solvent or with each other may be affected by temperature; and (c) the relative population of a conformational equilibrium may be altered by temperature.

Although the temperature dependence is rather small, it can be significant, and measurements at very low temperature have revealed conformational and stereochemical details. An example of such dependence was shown by Baghenov and Volkenshtein (1954), who found that methylvinylcarbinol (1) in hexane at 334 nm underwent a substantial decrease in optical rotation when the temperature was changed from 15°C ([α] + 195.0°) to 55°C ([α] + 93.7°). The solvent made little difference in this temperature effect on the alcohol (**1**). For accurate mea-

$$H_3C-C \overset{\displaystyle CH=CH_2}{\underset{\underset{H}{|}}{\diagdown}} OH$$

1

surement of some compounds such as tartaric acid derivatives, precise control of the temperature is required. Data obtained by these careful measurements can then be used to calculate the enthalpy and entropy differences in studies of the rates of racemization, as shown by Cagle and Eyring (1951), and in some equilibrium studies by Kauzmann *et al.* (1940). Many studies have been carried out with low-temperature measurements using optical rotatory dispersion and circular dichroism for conformational analysis, as reviewed by Djerassi (1960), Snatzke (1967), and Crabbé (1965).

B. Variation of Optical Rotation with Concentration

Conversion of Eq. (1) to the form for a solution gives

$$[\alpha]_\lambda^t = 100\alpha/lc, \tag{2}$$

where c is the concentration in grams per 100 ml of solution. The specific rotation $[\alpha]$ is not, of course, a CGS unit and is not always proportional to the concentration. Interaction of solute molecules may lead to nonlinear rotations for concentrated solutions. As pointed out by Heller and Curmè (1972), the specific rotation even in very dilute solution may not be truly constant and should be replaced by the term *intrinsic rotation* defined as

$$[\alpha]_{c\to0} - \{\alpha\}. \tag{3}$$

These authors showed, for example, that nicotine in ethanol has a specific rotation in a 5% solution differing by 0.8% from the intrinsic rotation, whereas in ethylene bromide the difference between $[\alpha]$ and $\{\alpha\}$ is more than 5%. In some extreme cases a change in sign of the rotation may occur, and these changes become more pronounced at wavelengths close to the optically active absorption bands. This is a special problem of which the investigator should be aware when calculating optical purity based on rotation data. It is essential to use the same concentrations when comparing solutions on an absolute basis. When the optical rotation of a new compound is reported, at least two different determinations should be made at different concentrations. This will indicate whether the optical rotation is concentration dependent. If it is, consideration should be given to handling the data by the intrinsic rotation method. Information can be obtained from the reference work by Heller and Curmè (1972) and from references therein.

C. Variation of Optical Rotation with Wavelength

The identification of the wavelength of measurement is essential because of the large change in optical rotation with wavelength, especially for substances having chromophores that absorb light in the visible or UV region and produce anomalous optical rotatory dispersion curves. These so-called Cotton effects may have points of enormous rotations, as in the case of the bridged biaryl **2**, which has a specific rotation at 310 nm of almost 25,000° (Mislow and Djerassi, 1960).

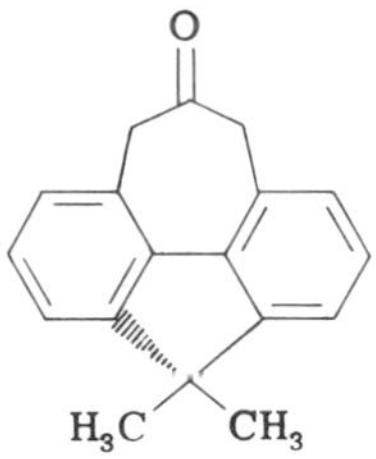

2

The D-line rotation was chosen as the standard wavelength for determinations as a result of the development of the Bunsen burner at a time when monochromatic light was especially needed for polarimetry. The ease of obtaining the yellow sodium flame by adding salts to the burner gave the sodium lines at 589.0 and 589.6 nm, which are essentially monochromatic. The alternative green mercury line at 546.1 nm has been used for natural-product studies and is a much better choice because of the substantially brighter light. It would be desirable to discontinue the use of the D-line, but it seems unlikely in view of the large number of instruments adapted to it and the voluminous data in the literature. The application of lasers as a light source is an area of future development.

In the measurement of rotation any path length may be used, but the establishment of the decimeter length by Biot (1838) arose from the need to obtain a reasonably large observed rotation. Similarly, the use of percent composition arose because molecular weights were uncertain at the time of Biot. Hence, the specific rotation can be converted to molecular rotation, designated as [φ] or [M] (Djerassi and Klyne, 1957), as shown in

$$[\phi]^t_\lambda = \frac{M}{100}[\alpha],\tag{4}$$

where M is the molecular weight. This term is used especially in reporting data for optical rotatory dispersion.

D. Variation of Optical Rotation with Solvent

The fact that a solute may interact with a solvent is well understood, but it is frequently forgotten in descriptions of the optical rotatory power of solutes. Since solutes may be affected by solvents, leading to complex molecular interactions, specific conformational changes, and variations in ionic species, it is obvious that these changes will affect optical rotations and produce appreciably different values for rotations depending on the solvent and/or the pH of the solution. Many amphoteric substances such as amino acids show a change in sign of the optical rotation when the pH of the solution is changed. This is expected because the specific rotations are often small ($<100°$), and their rotatory dispersion curves frequently cross the zero line at wavelengths between 400 and 500 nm. Thus, the change in molecular species from the zwitterion form to the amine salt of the undissociated carboxylic acid would have a profound effect on the rotation. Even such minor solvent changes as a shift from 40 to 60% dioxane in water was shown by Tanford (1962) to cause a shift in the D-line rotation of *N*-acetyl-L-glutamic acid from levorotatory to dextrorotatory. In addition, Greenstein and Winitz (1961) remarked that almost all L-amino acids undergo a shift in optical rotation to more positive values when the solvent changes from water to more acidic media.

The choice of a solvent for an unknown, potentially chiral molecule should be based on the observed solubility during the isolation and the potential interactions observed in the spectral data. It is desirable to record the UV–visible absorption spectrum to ascertain the possible presence of a chromophore in or near the visible region and to recognize that the concentration used for the spectral curve will generally be much smaller than that used for the rotation. Meaurement of the rotation in at least three different solvents or at different pH values will assist in finding the best conditions under which to obtain the larger rotations. In general, the temperature should be as low as practical, but many chemists prefer 25°C because that is the approximate temperature of most buildings. Ideally, the first choice of solvent is the most nonpolar solvent in which the product is soluble. Acids and bases should be measured in neutral, acidic, and basic solutions. Methanol is usually preferable to ethanol because of the smaller amount of water present and the ease of recovering the product. Spectral-grade solvents are best but are not essential for measurements at long wavelength.

A particular solvent may be used for one class of compounds because of the convenience of comparing the data with many known compounds. Thus, chloroform has been the primary choice for steroids and terpenes, and water the choice for carbohydrates, whereas aqueous acid solution or ethanol is commonly used for amines. Hydrocarbons of low molar mass offer some difficulties as solvents because of their volatility. Most solvents are transparent throughout the visible spectrum but, for rotatory dispersion or circular dichroism data, lack of absorption is essential, especially for measurements below 300 nm.

Problems may arise from an unexpected interaction of the chiral solute with the solvent. Preliminary UV–visible spectra may reveal these problems, but the unpredictability of the extinction coefficients may fail to give warning. Djerassi *et al.,* (1959) explored the extent of hemiketal formation on acidification of an alcoholic solution of steroids and other chiral ketones. The reduction of the rotation in the region of the Cotton effect band due to hemiketal formation was substantial when methanol was used as solvent but was minimal with isopropyl alcohol. The effect was observed even at the sodium D-line, where (+)-3-methylcyclohexanone showed a decrease in specific rotation in ethanol from +12 to 0° on acidification. That this is not 100% formation of the hemiketal was shown by the measurement of rotation at 310 nm, where the ethanol solution gave [α] +745° and the acidified solution decreased to +492°, indicating 33% formation of the hemiketal. From similar data for steroids, these authors used the variation in rotation with change in solvent to elucidate structural features of the steroid ketones.

The interaction of optically active solvents with racemic or partially resolved solutes has been used for the determination of absolute configuration by means of NMR spectroscopy (Pirkle and Beare, 1967). The fact that the interactions cause anomalous splitting patterns supports the need for using care in selecting a racemic solvent for the study of an optically active compound.

E. *Variation of Optical Rotation Due to Contaminants*

The reproducibility of rotation data may be destroyed by the presence in the solution of an unknown, apparently inactive contaminant. If the contaminant is carried with the chiral product, it would be expected to decrease the optical rotation, and further purification should identify and eliminate it. In some cases, however, such contaminants may enhance the rotatory power, leading to the erroneous conclusion that the sample is more pure than it actually is. One such example, reported by MacLeod *et al.* (1964), involved Ucon Polar Oil (Union Carbide Co.), which appeared in the sample after the presumed purification by gas chromatography. This was proved by the ultimate separation of the contaminant and measurement of the optical rotation of 66% optically pure (−)-*n*-butylphenylcarbinol with various amounts of optically inactive polyethylene glycol 300 (Carbowax), which showed larger rotations (up to 8% exaltation by 2.6% Carbowax). The origin of the exaltation of rotation has not been identified.

Yamaguchi and Mosher (1973) discovered an enhancement of optical rotation due to traces of acetophenone in the carbinol obtained by asymmetric reduction of the achiral ketone. This appears to result from induced dissymmetry of the ketone, probably through the formation of a hemiketal with the chiral alcohol or hydrogen bonding between the alcohol and ketone. The rotation at the D-line was

increased from $+43.1$ to $+58.3°$ by the presence of a 4-molar excess of acetophenone.

The enhancement of rotation by the addition of racemic compounds to a
solution of an optically active compound has been explored by a number of
investigators, who call this the Pfeiffer effect (Schipper, 1975). It is an area
under investigation (Gillard, 1979) by theoreticians who have proposed that the
enhancement of rotation is due to covalent bond formation when aromatic amines
are the added agent. Such an explanation is not valid in the case of Carbowax
cited above (MacLeod *et al.*, 1964). Other examples are known in which induced
chirality results because of the addition of a new longer-wavelength chromophore associated with an achiral compound (Lyle *et al.*, 1979). This means
that the rotation should increase because the Cotton effect due to the induction of
dissymmetry in the achiral substrate will appear at longer wavelength (especially
in cases involving aromatic amines). The induction of dissymmetry does not
require covalent bond formation but can arise through an induced chiral conformation between the substrate and added agent. An example of a compound in
which the conformation could be stabilized by a hydrogen bond is **3**; here, one or
more chiral alcohols could interact with the racemic nitrosamines.

R*OH = optically active sugar or alcohol

3

The problem of obtaining any optical rotatory power with certain types of
chiral molecules has been considered by Wynberg and Hulshof (1974). Tetra-
alkylmethanes are expected to have low optical rotation (Thomson, 1953) because of the similar polarizability of the four alkyl groups. The combination of
extremely low rotations and an absorption maximum of wavelength only in the
far UV region which cannot enhance the rotation at the D-line complicates the
observation of any rotatory power. Hoeve and Wynberg (1980) have offered
evidence that butylethylhexylpropylmethane (**4**) has zero optical rotation between 280 and 580 nm, although the compound is 25% optically pure, as determined by several methods. Calculation of its optical rotation gave a value of
about 1×10^{-4} degree at 365 nm, which is below the noise level of current
instruments.

4

As the measurement of optical rotation becomes more precise as a result of increased instrumental sensitivity, the problem of duplicating experimental measurements will increase. Rotations cited in the literature must be carefully evaluated before the status of enantiomeric purity is granted to any compound, because the potential for error in obtaining an optical rotation is significant, especially when the choice of conditions can cause variations such as those cited above. The fact that the enantiomeric composition can be determined by several different methods eliminates the necessity of obtaining optical rotations for compounds of low rotatory power or minimal enantiomeric purity. In the great majority of cases, however, the most useful reference property for an optically active compound is the optical rotation, preferably measured at several wavelengths and in at least two solvents. The designation of enantiomeric composition of a compound on the basis of its optical rotation should be confirmed by a second method before the absolute rotation is assigned to the enantiomers. Because in the majority of cases the optical purity, based on rotation, and the enantiomeric excess, based on other physical methods of determination, are identical, the more convenient experimental procedure of optical rotation can be used for studying asymmetric synthesis. If precise methods are necessary, however, the equivalency of these descriptions must be established.[2]

References

Baghenov, N. M., and Volkenshtein, M. (1954). *Zh. Fiz. Khim.* **28,** 1299.

Barton, D. H. R. (1945). *J. Chem. Soc.,* p. 813.

Biot, J. B. (1838). *Mem. Acad. Sci. Inst. Fr.* 15, 93.

Bláha, K., Červinka, O., and Kovář, J. (1971). "Fundamentals of Stereochemistry and Conformational Analysis." Illiffe, Prague.

Brewster, J. H. (1959). *J. Am. Chem. Soc.* **81,** 5475, 5483, 5493.

Cagle, F. W., Jr., and Eyring, H. (1951). *J. Am. Chem. Soc.* **73,** 5628.

Crabbé, P. (1965). "Optical Rotatory Dispersion and Circular Dichroism in Organic Chemistry." Holden-Day, San Francisco.

Djerassi, C. (1960). "Optical Rotatory Dispersion." McGraw-Hill, New York.

Djerassi, C., and Klyne, W. (1957). *Proc. Chem. Soc. London,* p. 55.

Djerassi, C., Mitscher, L. A., and Mitscher, B. J. (1959). *J. Am. Chem. Soc.* **81,** 947.

Elsenbaumer, R. L., and Mosher, H. S. (1979). *J. Org. Chem.* **44,** 600.

Freudenberg, K., Kuhn, W., and Bumann, I. (1930). *Chem. Ber.* **63,** 2380.

---

[2]In addition to sharing with the other contributors in honoring Professor Harry Mosher, to whom this volume is dedicated, the authors of this chapter dedicate their contribution to Dr. Ulrich Weiss in the year of his seventy-fifth birthday (24 January 1983). Dr. Weiss has made a great contribution to chemistry with his studies of optical activity, especially the optical rotatory dispersion and circular dichroism of alkaloids, terpenes, and diene systems.

Gillard, R. D. (1979). *In* "Optical Activity and Chiral Discrimination" (S. F. Mason, ed.), pp. 353–367. Reidel, Boston.

Greenstein, J. P., and Winitz, M. (1961). "Chemistry of the Aminoacids," Vol. 1. Wiley, New York.

Heller, W., and Curmè, H. G. (1972). *In* "Physical Methods of Chemistry" (A. Weissberger and B. W. Rossiter, eds.), Part IIIC, pp. 51–181. Wiley (Interscience), New York.

Hoeve, W. T., and Wynberg, H. (1980). *J. Org. Chem.* **45**, 2754.

Horeau, A. (1969). *Tetrahedron Lett.,* p. 3121.

Horeau, A. (1977). *In* "Stereochemistry, Fundamentals and Methods" (H. B. Kagan, ed.), Vol. 3, pp. 51–94. Thieme, Stuttgart.

Hudson, C. S. (1909). *J. Am. Chem. Soc.* **39**, 66.

Jacques, J., Collet, A., and Wilen, S. H. (1981). "Enantiomers, Racemates, and Resolutions." Wiley (Interscience), New York.

Kauzmann, W. J., Walter, J. E., and Eyring, H. (1940). *Chem. Rev.* **26**, 339.

Klyne, W. (1954). *Chem. Ind. (London),* p. 1198.

Klyne, W., and Buckingham, J. (1978). "Atlas of Stereochemistry," 2nd ed. Oxford Univ. Press, New York.

Lowry, T. M. (1935). "Optical Rotatory Power." Longmans, Green, London; 2nd ed. (1964), Dover, New York.

Lyle, R. E., Fribush, H. M., Singh, S., Saavedra, J. E., Lyle, G. G., Barton, R., Yoder, S., and Jacobson, M. K. (1979). *ACS Symp. Ser.,* No. 101.

MacLeod, R., Prosser, H., Fikentscher, L., Lanyi, J., and Mosher, H. S. (1964). *Biochemistry* **3**, 838.

Mills, J. A. (1952). *J. Chem. Soc.,* p. 4976.

Mislow, K., and Djerassi, C. (1960). *J. Am. Chem. Soc.* **82**, 5247.

Pirkle, W. H., and Beare, S. D. (1967). *J. Am. Chem. Soc.* **89**, 5485.

Plattner, Pl. A., and Heusser, H. (1944). *Helv. Chim. Acta* **27**, 748.

Schipper, P. E. (1975). *Inorg. Chim. Acta* **12**, 199.

Schwarz, J. C. P. (1964). *In* "Physical Methods in Organic Chemistry" (J. C. P. Schwarz, ed.), pp. 235–243. Holden-Day, San Francisco.

Snatzke, G. (1967). *In* "Optical Rotatory Dispersion and Circular Dichroism in Organic Chemistry" (G. Snatzke, ed.), pp. 335–340. Heyden & Son, London.

Tanford, C. (1962). *J. Am. Chem. Soc.* **84**, 1747.

Thomson, T. R. (1953). *J. Am. Chem. Soc.* **75**, 6070.

Wynberg, H., and Hulshof, L. A. (1974). *Tetrahedron* **30**, 1775.

Yamaguchi, S., and Mosher, H. S. (1973). *J. Org. Chem.* **38**, 1870.

3

Competitive Reaction Methods for the Determination of Maximum Specific Rotations

Alain R. Schoofs
Centre de Recherche Delalande
Rueil-Malmaison, France

Jean-Paul Guetté
Laboratoire de Chimie Organique
Conservatoire National des Arts et Métiers
Paris, France

I. Introduction

Until 1959 there was no general method for the certain determination of the specific rotation of a pure enantiomer. Aside from a few cases, it was therefore impossible to determine with certitude the purity of enantiomers obtained in an asymmetric synthesis or a resolution. Since the 1960s there have been methods for the determination of the enantiomeric composition of a partially resolved mixture. These methods, whether enzymatic, radiochemical, spectrometric, chromatographic, or calorimetric, have been constantly improved, and they now possess great versatility (Jacques *et al.*, 1981).

Before such analytical methods were developed many authors attempted to find other means of calculating the specific rotation of a pure enantiomer. We discuss here various methods using kinetic resolution, which can be grouped as (*a*) methods that involve an asymmetric destruction of a racemic mixture and (*b*) methods that are based on two reciprocal kinetic resolutions (Horeau's method and related variations).

II. Principles of Kinetic Resolution

When a racemic compound $A_{\pm}$ is transformed by means of a chiral reactant C_{+}, the rates of transformation of the enantiomers A_{+} and A_{-} can be different. This situation corresponds to two competitive reactions for which the rate constants are, respectively, k_p and k_N:

$$A_{+} + C_{+} \xrightarrow{k_p} A'_{+}$$

$$A_{-} + C_{+} \xrightarrow{k_N} A'_{-}$$

The difference between the two rate constants depends on the difference between the free energy of activation of the two competing reactions (Fig. 1).

If the transformation of $A_{\pm}$ by C_{+} is interrupted before its completion, the remaining portion of A will be more or less resolved depending on the k_p/k_N ratio and on the extent of the reaction. Several authors have tried to quantify this phenomenon and to explore all of its possible applications.

III. Calculation of the Specific Rotation of an Enantiomer by Means of a Method Using Asymmetric Destruction of the Corresponding Racemic Mixture

Early in the twentieth century Fajans (Bredig and Fajans, 1908; Fajans, 1910) established the law of variation of the enantiomeric purity P_A of the remaining compound A in relation to its enantioselective destruction by an enzymatic system. However, it is in the area of nonenzymatic asymmetric decomposition that the most progress has been made.

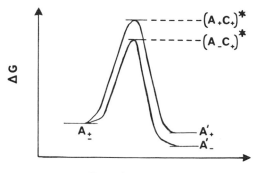

Reaction coordinate

Fig. 1. Free energy reaction coordinate diagram.

The first detailed study was accomplished by Kuhn and Knopf (1930). In the course of a famous experiment these authors, working from a suggestion made by LeBel (1874), showed that circularly polarized light can act as a chiral reactant, causing enantioselective photodecomposition of racemic α-azido-*N,N*-dimethylpropionamide ($1_\pm$).

$$1_\pm \quad + \quad \text{achiral products} \quad + \quad N_2$$

When the extent of the reaction was 50%, the product (**1**) that remained had an enantiomeric purity of 0.5%. A kinetic resolution of $1_\pm$ was therefore obtained.

The stereoselectivity of this asymmetric destruction depends on the ratio of the rate constants k_p and k_N, which are themselves proportional to the molecular absorption coefficients ϵ_+ and ϵ_- of the enantiomers A_+ and A_-.

With the anisotropy factor g, which is defined by

$$g = \frac{\Delta\epsilon}{\epsilon} = \frac{\epsilon_+ - \epsilon_-}{(\epsilon_+ + \epsilon_-)/2} = 2\frac{\epsilon_+ - \epsilon_-}{\epsilon_+ + \epsilon_-},$$

one can calculate the ratio of the rate constants

$$K = k_p/k_N = (2 + g)/(2 - g).$$

Because the asymmetric photodecomposition reactions are first order, the rates of disappearance of each enantiomer at any given time t are

$$d[A_+]/dt = -k_p[A_+]; \qquad d[A_-]/dt = -k_N[A_-].$$

If $[A_+]_0$ and $[A_-]_0$ are the initial concentrations of each enantiomer ($t = 0$), the new equations

$$[A_+] = [A_+]_0 e^{-k_p t} \tag{1}$$

and

$$[A_-] = [A_-]_0 e^{-k_N t} \tag{2}$$

are obtained by integration.

At any moment the enantiomeric purity P_A of A is defined by

$$P_A = \frac{[A_+] - [A_-]}{[A_+] + [A_-]} = \frac{[A_+]_0 e^{-k_p t} - [A_-]_0 e^{-k_N t}}{[A_+]_0 e^{-k_p t} + [A_-]_0 e^{-k_N t}}.$$

Because for $t = 0$, $[A_+]_0 = [A_-]_0$, the enantiomeric purity is established by

$$P_A = \frac{\exp[(k_N - k_p)t] - 1}{\exp[(k_N - k_p)t] + 1} = \tan h \tfrac{1}{2}(k_N - k_p)t. \tag{3}$$

The curve in Fig. 2, a graphic representation of this important relation, shows that, whatever the ratio $K = k_p/k_N$, the enantiomeric purity P_A tends toward 1 when the time t tends toward infinity.

If it is possible to determine the values of k_p and k_N experimentally, the relationship expressed in Eq. (3) allows one to calculate the specific rotation of the pure enantiomer A [the enantiomeric purity and the optical purity are equivalent (Horeau, 1969)].

In 1974 Balavoine et al. reexamined and extended earlier work. They succeeded in photodecomposing racemic camphor asymmetrically. With an extent of photodestruction of 99%, they recovered partially resolved camphor with an enantomeric purity of 20%.

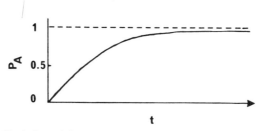

Kagan et al. have reconsidered the relationships established by Kuhn and Knopf (1930), and they have proposed a new approach that emphasizes the importance of the ratio k_p/k_N during the course of a kinetic resolution. If x represents the extent of the reaction, then

Fig. 2. Variation of the enantiomeric purity P_A as a function of time.

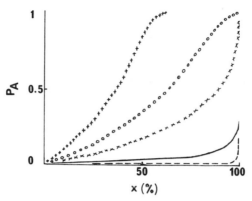

Fig. 3. Relation between enantiomeric purity P_A and extent of the reaction x as a function of the ratio $K = k_p/k_N$. Key: +, $K = 39$; o, $K = 3$; ×, $K = 1.67$; —, $K = 1.11$; ---, $K = 1.02$.

$$x = \frac{([A_+]_0 + [A_-]_0) - ([A_+] + [A_-])}{[A_+]_0 + [A_-]_0} = 1 - \frac{[A_+] + [A_-]}{[A_+]_0 + [A_-]_0}.$$

By the use of Eqs. (1) and (3), the following conclusion can be reached:

$$x = 1 - \frac{e^{-k_p t} + e^{-k_N t}}{1 + 1} = 1 - \frac{e^{-k_p t} + e^{-k_N t}}{2}. \tag{4}$$

By eliminating t from Eqs. (3) and (4), one can determine the variation of the enantiomeric purity P_A as a function of the extent of the reaction for a given value of $K = k_p/k_N$ (and see Fig. 3):

$$x = 1 - \tfrac{1}{2}\left[\frac{(1 + P_A)^{1/(1-K)}}{1 - P_A} + \frac{(1 + P_A)^{K/(1-K)}}{1 - P_A}\right]. \tag{5}$$

The results of two kinetic resolutions performed with different extents of conversion, x_1 and x_2, allow one to calculate the specific rotation of enantiomerically pure A.

It must be emphasized that Eqs. (1)–(5) are of general applicability to every chemical system leading to kinetic resolution if the experimental conditions are appropriate. Thus, in a classic investigation of the absolute configuration of 1′,4″-dinitrobenzo-1,2,3,4-cycloheptadi-1,3-en-6-one (DNBCH-6-one), Mislow and co-workers (Newman *et al.*, 1958) studied quantitatively the asymmetric reduction of DNBCH-6-one using a chiral reducing complex formed by (S)-(+)-methyl-t-butylcarbinol and Al(OtBu)$_3$ (Fig. 4). The levorotatory enantiomer of the ketone is preferentially reduced. Thus, when the reduction of the racemic ketone is incomplete, a kinetic resolution results ($k_p \neq k_N$). The chiral reduction complex was used in great excess. The concentrations of A$_+$ and A$_-$ vary

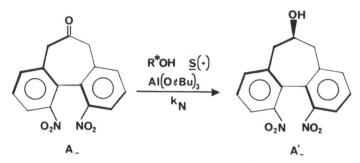

Fig. 4.   Representation of the enantiomer (A_) of DNBCH-6-one that is preferentially reduced.

according to Eqs. (1) and (2), where $[A]_0$ corresponds to the initial concentration of the racemic ketone:

$$[A_+] = \frac{[A]_0}{2}e^{-k_p t}; \qquad [A_-] = \frac{[A]_0}{2}e^{-k_N t}.$$

If P_A is the enantiomeric purity of the partially resolved ketone and $P_{A'}$ that of the resulting alcohol, it is possible to establish the new relationship

$$\frac{P_{A'}}{P_A} = \frac{[A_+] - [A_-]}{[A'_+] - [A'_-]} .$$

From these equations, the values P_A and $P_{A'}$ can be expressed as a function of time:

$$P_A = \frac{e^{-k_p t} - e^{-k_N t}}{e^{-k_p t} + e^{-k_N t}}, \tag{6}$$

$$P_{A'} = \frac{e^{-k_p t} - e^{-k_N t}}{2 - (e^{-k_p t} + e^{-k_N t})} . \tag{7}$$

In this manner Mislow and co-workers were able to calculate with great precision the specific rotation of the alcohol (A') and that of the ketone (A).

It is possible to establish another relationship between P_A and $P_{A'}$, taking into account the extent of the reaction x:

$$P_{A'} = \frac{([A_-]_0 - [A_-]) - ([A_+]_0 - [A_+])}{([A_-]_0 - [A_-]) + ([A_+]_0 - [A_+])},$$

$$= \frac{[A_+] - [A_-]}{([A_-]_0 + [A_+]_0) - ([A_+] + [A_-])},$$

and

$$P_{A'} = \frac{[A_+] - [A_-]}{[A_+] + [A_-]}\frac{1 - x}{x} = P_A \frac{1 - x}{x} . \tag{8}$$

If the yield of the reaction is 50% ($x = 0.5$), the enantiomeric purities P_A and $P_{A'}$ are equal.

Lamaty and co-workers (Conan et al., 1974) have studied the stereoselective hydrolysis of diastereomeric 1,3-dioxolanes formed by the reaction of (R,R)-diethyl tartrate with several chiral ketones. If the stereoselective hydrolysis is incomplete, the result of the two reactions corresponds to the kinetic resolution of the racemic ketone $A_\pm$ (Fig. 5).

$$A_+B_+ \xrightarrow[k_p]{H_2O} A_+ + B_+$$

$$A_-B_+ \xrightarrow[k_N]{H_2O} A_- + B_+$$

The enantiomeric purity of each partially resolved ketone A is represented by an equation identical to

$$P_A = \frac{[A_-] - [A_+]}{[A_-] + [A_+]} = \frac{e^{-k_N t} - e^{-k_p t}}{2 - (e^{-k_N t} + e^{-k_p t})}$$

The specific rotations $[\alpha]_0$ that are calculated by this method for various ketones are practically identical to experimental values appearing in the literature.

The enantioselective rearrangement of racemic 3-methyl- and 3-tert-butyl-indenes ($A_\pm$, R = Me and tBu) into achiral indenes (C) in the presence of an asymmetric catalyst B* (quinidine or hydroquinidine) has been analyzed in detail (Meurling and Bergson, 1974; Meurling et al., 1976).

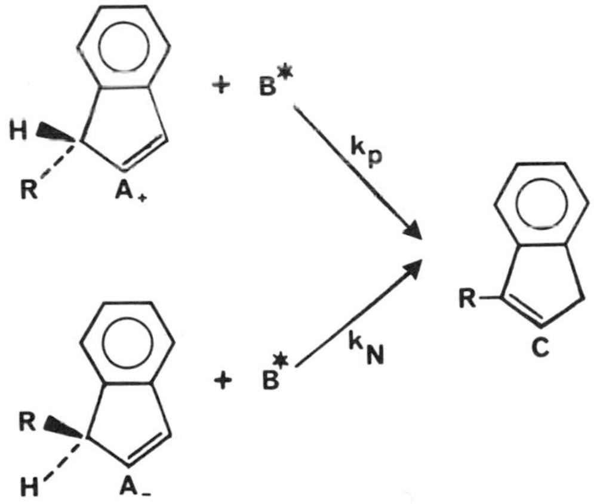

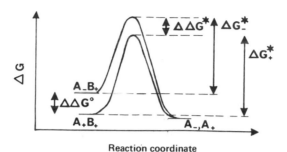

Fig. 5. Free energy reaction coordinate diagram.

Because two nonreversible, competitive, pseudo-first-order reactions are in question, the variation of the specific rotation of partially resolved starting material as a function of time is given by

$$[\alpha] = \frac{A_0}{2}[\alpha]_0(e^{-k_p t} - e^{-k_N t}),$$

where A_0 is the number of moles of racemic A and $[\alpha]_0$ the specific rotation of enantiomerically pure A.

The corresponding enantiomeric purity P_A is easily determined as a function of the extent of reaction x:

$$P_A = [\alpha]/[\alpha]_0 = 1/A_0(1 - x). \qquad (9)$$

In well-documented articles the limitations of methods based either on simple polarimetric measurements and chromatographic analysis of A and C or on mathematical models necessitating sophisticated numerical processes have been compared and discussed (Meurling and Bergson, 1974; Meurling et al., 1976). Other workers (Brandt et al., 1977) have numerically calculated the variations in respective concentrations of A_+ and A_- in the most general case of kinetic resolution of $A_\pm$ by the partially resolved mixture (B_+, B_-) as a function of the ratio of the various rate constants.

These theoretical considerations have been well illustrated (Martin et al., 1981). Allylic alcohols of high enantiomeric purity have been obtained by kinetic resolution during the asymmetric epoxidation of the corresponding racemic mixtures by the L-diisopropyl tartrate (DIPT)/titanium(IV) isopropoxide [Ti(OiPr)₄]/ tert-butyl hydroperoxide (TBHP) system (Fig. 6.).

The ratio of the rate constants for the enantiomers studied varied between 15 and 140. Even at epoxidation yields of 55 ± 5% unreacted alcohols of very high enantiomeric purity were recovered. Excellent agreement between experimental

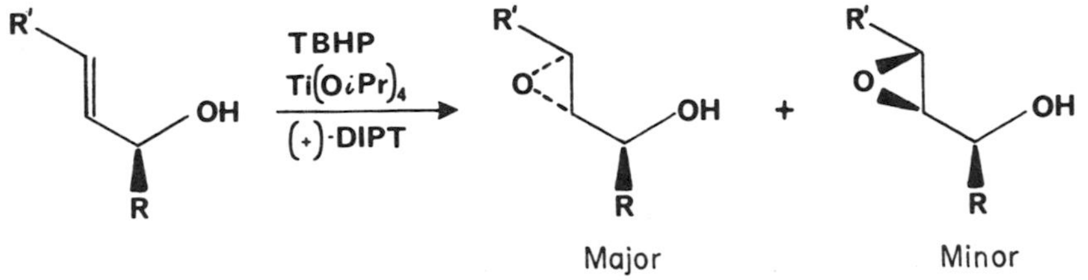

Fig. 6. Representation of the conversion of the most rapidly epoxidized allylic alcohol to diastereomeric products.

enantiomeric purities and those calculated with the aforementioned formulas was observed.

IV. Calculation of the Specific Rotation of an Enantiomer by Means of a Method Using Two Reciprocal Kinetic Resolutions

Horeau (1961, 1977) has discussed the kinetic resolution of racemic α-phenylbutyric acid as a means of establishing the absolute configuration of chiral secondary alcohols. Horeau has also proposed an original method for calculating the specific rotation of a pure enantiomer based on the principle of kinetic resolution already described (Horeau, 1964a).

If a racemate, $A_{\pm}$, is allowed to react with a pure enantiomer, say B_+, the transition states for the reaction of the enantiomers A_+ and A_- with B_+ are diastereomeric, and they are located at different energy levels on an energy diagram. The rate of reaction of A_+ with B_+ is different from that of A_- with B_+, and the two rate constants, k_p and k_N, of these two reactions are then unequal:

$$A_+ + B_+ \xrightarrow{k_p} A_+ B_+ \tag{10}$$

$$A_- + B_+ \xrightarrow{k_N} A_- B_+ \tag{11}$$

For reasons of symmetry it is also possible to write

$$A_+ + B_- \xrightarrow{k_N} A_+ B_- \tag{12}$$

$$A_- + B_- \xrightarrow{k_p} A_- B_- \tag{13}$$

When a pure enantiomer, B_+, reacts with an excess of a racemic chiral compound, $A_{\pm}$, the reaction can be described by

$$[A_+]_0 + [A_-]_0 + [B_+]_0 \rightarrow \begin{cases} \dfrac{[A_+B_+] + [A_-B_+]}{nra \qquad nr(1-a)} \\[2ex] \dfrac{[A_+] + [A_-]}{N - nra \quad N - nr(1-a)} \end{cases} \qquad (14)$$

where $2N$ is the number of moles of A, n is the number of moles of B_+, $2N \gg n$, r is the chemical yield of the reaction, and a is the molar ratio of B_+ that has reacted with A_+. If k_p is different from k_N [Eqs. (10)–(13)], a is different from 0.5 and the quantities of A_+ and A_- remaining at the end of the reaction are unequal:

For $k_p > k_N$ $a > 0.5$ $[A_-] > [A_+]$

For $k_p < k_N$ $a < 0.5$ $[A_-] < [A_+]$

Such a system permits the kinetic resolution of $A_\pm$ by B_+.

Marckwald and McKenzie (1899) clearly noted the possibilities offered by the application of this principle many years ago. They observed that the heating of racemic mandelic acid with natural menthol at 155°C for 1 h resulted in partial esterification. The unreacted mandelic acid had a specific rotation of $-3.3°$, corresponding to a very low optical purity ($\sim 2\%$). Following up this early lead, Horeau found chiral auxiliary compounds that greatly improved the efficiency of kinetic resolution, thus opening the way for many interesting applications. It is possible to establish the enantiomeric purity of A at the end of the kinetic resolution:

$$P_A = \frac{[A_-] - [A_+]}{[A_-] + [A_+]} = \frac{[N - nr(1-a)] - [N - nra]}{[N - nr(1-a)] + [N - nra]}$$

$$= (2a - 1)\frac{nr}{2N - nr}. \qquad (15)$$

Mislow (1966) defined the stereoselectivity Q of the reaction as the ratio of the difference between the quantities of diastereomers over their sum:

$$Q = \frac{[A_+B_+] - [A_-B_-]}{[A_+B_+] + [A_-B_-]} = \frac{nra - nr(1-a)}{nra + nr(1-a)} = 2a - 1. \qquad (16)$$

If the reaction is stereospecific, $a = 1$ and $Q = 1$, the enantiomeric purity of A obtained reaches a maximum value of

$$P_A = nr/(2N - nr).$$

If partially resolved B is used instead of enantiomerically pure B, the relationship expressed in Eq. (14) takes the form shown in Eq. (17), where P_B is the enantiomeric purity of B. The enantiomeric purity of A obtained by kinetic

$$\left\{ \begin{array}{c} \begin{array}{ccc} [A_+B_+] & + & [A_-B_-] \\ nra\left(\dfrac{1+P_B}{2}\right) & & nra\left(\dfrac{1-P_B}{2}\right) \end{array} \\[2em] \begin{array}{ccc} [A_+B_-] & + & [A_-B_+] \\ nr(1-a)\left(\dfrac{1-P_B}{2}\right) & & nr(1-a)\left(\dfrac{1+P_B}{2}\right) \end{array} \\[2em] \begin{array}{c} + \ [A_+] \\ N - nra\left(\dfrac{1+P_B}{2}\right) - nr(1-a)\left(\dfrac{1-P_B}{2}\right) \end{array} \\[2em] \begin{array}{c} + \ [A_-] \\ N - nra\left(\dfrac{1-P_B}{2}\right) - nr(1-a)\left(\dfrac{1+P_B}{2}\right) \end{array} \end{array} \right. \tag{17}$$

$$\begin{array}{cc} [A_+]_0 & + & [A_-]_0 \\ N & & N \\[1em] [B_+]_0 & + & [B_-]_0 \\ n\left(\dfrac{1+P_B}{2}\right) & n\left(\dfrac{1-P_B}{2}\right) \end{array} \xrightarrow{}$$

resolution (P_A) can then be calculated using

$$P_A = \frac{[A_-] - [A_+]}{[A_-] + [A_+]} = (2a - 1)\frac{nr}{2N - nr}P_B. \tag{18}$$

It becomes clear that the enantiomeric purity of A is directly proportional to the enantiomeric purity of the compound B used. The stereoselectivity of the reaction, as previously defined, is

$$Q = \frac{([A_+B_+] + [A_-B_-]) - ([A_+B_-] + [A_-B_+])}{([A_+B_+] + [A_-B_-]) + ([A_+B_-] + [A_-B_+])} . \tag{19}$$

We can calculate Q with the help of

$$Q = 2a - 1 = \frac{2N - nr}{nr}\frac{P_A}{P_B} , \tag{20}$$

Of course, if compounds A and B are engaged in the reaction in racemic form, the result will be

$$P_A = P_B = 0$$

The relationship shown in Eq. (20) leads to an indetermination that can be eliminated if this is taken into account. The relation defining Q therefore becomes

$$Q = 2a - 1.$$

This relationship is very important. It shows, in fact, that it is possible to determine the stereoselectivity Q of the kinetic resolution of $A_{\pm}$ with B_+ (or B_-) not only using enantiomerically pure B_+ [Eq. (16)], but by using racemic B as well. In the latter case kinetic resolution of $A_{\pm}$ by $B_{\pm}$ will not be possible, but it will be possible to calculate Q by chromatographic or spectrometric determination of the racemic diastereomers formed. This interesting property allows for the facile determination of the stereoselectivity of the interaction of two chiral compounds without necessitating their resolution and avoiding all risk of racemization of the resolved partners. It is important to note that the value of Q is independent of the proportions of A and B, as well as the chemical yield of the reaction. This is true not only when B is racemic, but also when B is enantiomerically pure.

If, in the course of a second kinetic resolution, n' molecules of enantiomerically pure A_- are allowed to react with $2N'$ molecules of racemic B_+, relationships analogous to those established for the reaction of $A_\pm$ with B_+, are obtained [Eq. (21)]. The quantity a' represents the mole fraction of A_- that combines with B_-, and r' is the chemical yield of the reaction.

$$
\begin{array}{ccc}
[B_+]_0 + [B_-]_0 + [A_-]_0 \\
N' \qquad N' \qquad n'
\end{array}
\rightarrow
\left\{
\begin{array}{cc}
[A_-B_-] & + & [A_-B_+] \\
n'r'a' & & n'r'(1-a) \\[2mm]
[B_-] & + & [B_+] \\
N' - n'r'a' & N' - n'r'(1-a')
\end{array}
\right.
\tag{21}
$$

The stereoselectivity Q' of the reaction is given by

$$
Q' = \frac{[A_-B_-] - [A_-B_+]}{[A_-B_-] + [A_-B_+]} = 2a' - 1. \tag{22}
$$

The enantiomeric purity of the isolated product B_+ is

$$
P'_B = (2a' - 1)\frac{n'r'}{2N' - n'r'}. \tag{23}
$$

If the product A is not enantiomerically pure, the enantiomeric purity of the product B that is obtained becomes

$$
P'_B = (2a' - 1)\frac{n'r'}{2N' - n'r'}P'_A, \tag{24}
$$

and the stereoselectivity Q' of the reaction is expressed by

$$
Q' = 2a' - 1 = \frac{2N' - n'r'}{n'r'}\frac{P'_B}{P'_A}. \tag{25}
$$

It is possible to relate the stereoselectivities Q and Q' of the two aforementioned kinetic resolutions by taking into account the symmetry relationship already described and supposing that the corresponding reactions are first order in each of the constituents. Thus, during the course of the kinetic resolution of $A_\pm$ by B_+, the rates of formation of the two diastereoisomers A_+B_+ and A_-B_+ are, respectively,

$$
V_{A_+B_+} = k_p[A_+][B_+]
$$

and

$$
V_{A_-B_+} = k_N[A_-][B_+].
$$

From this it can be deduced that

$$\frac{[A_+B_+]}{[A_-B_+]} = \frac{V_{A_+B_+}}{V_{A_-B_+}} = \frac{k_p[A_+]}{k_N[A_-]}.$$

If a large excess of $A_\pm$ over B_+ is used and if the stereoselectivity of the reaction is not very high, the following relations are valid throughout the reaction:

$$[A_+] \neq [A_-] \quad \text{and} \quad \frac{[A_+B_+]}{[A_-B_+]} = \frac{k_p}{k_N}.$$

This allows for the calculation of Q:

$$Q = \frac{[A_+B_+] - [A_-B_+]}{[A_+B_+] + [A_-B_+]} = \frac{1 - (k_N/k_p)}{1 + (k_N/k_p)} = \frac{k_p - k_N}{k_p + k_N}.$$

Analogously, in the case of the kinetic resolution of $B_\pm$ by A_-, it is also possible to write

$$V_{A_-B_-} = k_p[A_-][B_-]$$

and

$$V_{A_-B_+} = k_N[A_-][B_+].$$

If $B_\pm$ is utilized in great excess over A_-,

$$\frac{[A_-B_+]}{[A_-B_-]} = \frac{V_{A_-B_+}}{V_{A_-B_-}} = \frac{k_N}{k_p},$$

which allows for calculation of the stereoselectivity Q':

$$Q' = \frac{[A_-B_-] - [A_-B_+]}{[A_-B_-] + [A_-B_+]} = \frac{k_p - k_N}{k_p + k_N}.$$

Under these conditions, the stereoselectivities Q and Q' of the two kinetic resolutions are thus identical:

$$Q = Q' = (k_p - k_N)/(k_p + k_N).$$

Equations (20) and (25), corresponding to kinetic resolutions in which A and B are not enantiomerically pure, give rise to

$$\frac{2N - nr}{nr}\frac{P_A}{P_B} = \frac{2N' - n'r'}{n'r'}\frac{P_B}{P'_A}. \tag{26}$$

The hypotheses that have been used for purposes of calculation having been taken into account, the general equation [Eq. (26)] relating the results obtained for the case of two kinetic resolutions is valid provided that one of the reactants is used in large excess and the chemical yield and the stereoselectivity of each resolution are not too high. Furthermore, the two kinetic resolutions must be "reciprocal." It has been shown that this reciprocity is not realized if an intermediary compound is formed (King and Sim, 1973; King et al., 1973). The kinetic laws are then not as simple and the preceding calculations are no longer possible.

In the case of reciprocal resolutions, the relationship expressed in Eq. (26) allows for the calculation of the specific rotation of A_+ if that of B_+ is known, or vice versa. For example, if the specific rotation of A_+ is known, at the end of the first resolution the enantiomeric purity of the isolated product A will be

$$P_A = [\alpha]_A/[\alpha_0]_A.$$

That of B will be

$$P_B = [\alpha]_B/[\alpha_0]_B.$$

At the end of the second resolution,

$$P'_A = [\alpha']_A/[\alpha_0]_A \quad \text{and} \quad P'_B = [\alpha']_B/[\alpha_0]_B.$$

Equation (26) then generates the following important equation:

$$[\alpha_0]_B^2 = [\alpha]_B[\alpha']_B \frac{[\alpha_0]_A^2}{[\alpha]_A[\alpha']_A} \frac{nr}{n'r'} \frac{2N' - n'r'}{2N - nr}. \tag{27}$$

It is therefore possible to calculate $[\alpha_0]_B$ by determining the chemical yield of each kinetic resolution and by measuring the specific rotation of A and of B before and after each resolution.

Horeau (1964a,b) demonstrated the validity of this method by calculating the specific rotations of phenylisopropylcarbinol (3), phenyl-n-propylcarbinol (2), and amphetamine (5) as a result of two kinetic resolutions carried out using first racemic and then the dextrorotatory anhydride of α-phenylbutyric acid (6). The values of the specific rotations calculated in each case agree with the experimental values in the literature. In the case of phenyl-n-butylcarbinol (4) Horeau et al. (1966) showed that this method yields results comparable to those obtained by chromatographic analysis of diastereomeric derivatives. The previously published values of the specific rotation of this compound were thus corrected.

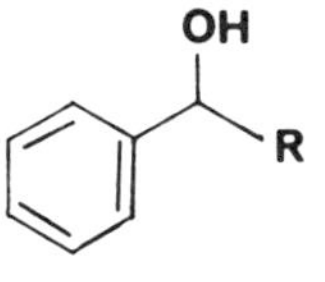

2 R = nPr
3 R = iPr
4 R = nBu

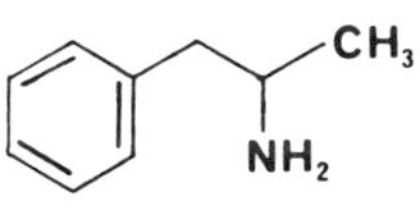

5

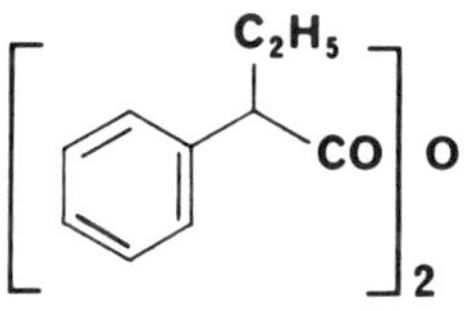

6

Christol *et al.* (1968) utilized Horeau's method for calculating the specific rotation of the spiro-4,4-nonyl-1-amine (**7**) using α-phenylbutyric acid anhydride (**6**) as auxiliary. Brugidou *et al.* (1974) showed that a variation of this method allows for the determination of the specific rotations of ketones and glycols such as **8** and **9**, the kinetic resolution being performed by partial ketalization.

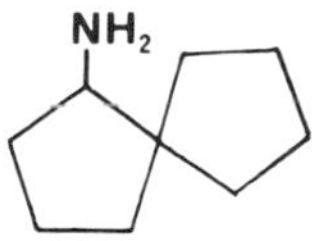

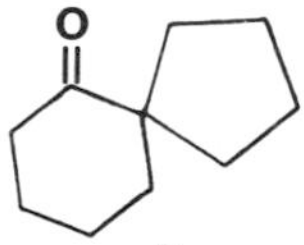

Horeau and colleagues have proposed several modifications of the original method. In the case of chiral secondary alcohols, Schoofs and Horeau (1977a) clearly demonstrated that it is possible to calculate enantiomeric purities that are less than 1%. On the same basis, these authors (Schoofs and Horeau, 1977b) also proposed a method for determining not only the enantiomeric purity, but also the absolute configuration of the same alcohols. Horeau (1975) showed that the kinetic resolution of a chiral compound makes it possible, under certain conditions, to obtain an enantiomer of this compound containing less than 0.1% of its antipode, which results in the determination of its specific rotation with this precision.

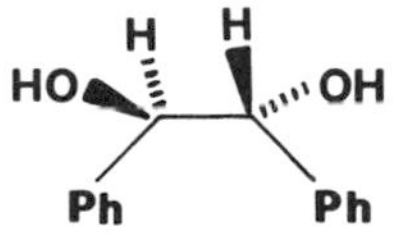

Horeau and Nouaille (1966) proposed a method for the simultaneous determination of the absolute configuration and the enantiomeric purity of 1-deutero-primary alcohols (**10**).

10

This method is also based on the principles of kinetic resolution. One enantiomer of chiral alcohol is esterified by each of the enantiomers of α-phenylbutyric acid (used in anhydride form) at different rates. The authors showed that the ratio of the rate constants is very nearly unity ($k_p/k_N = 1.0003$), and they assumed that this ratio does not depend on the nature of R. According to this assumption, the stereoselectivity of the reaction comes essentially from the difference in size between H and D, D being smaller than H (Horeau *et al.*, 1965). This hypothesis allows for the calculation of the specific rotations of chiral 1-deutero-primary alcohols, even if only small quantities of these alcohols (~100 mg) are available.

V. Conclusion

The results that have been summarized here demonstrate that kinetic resolutions constitute a viable method of calculating the specific rotation of pure enantiomers. It is also important to note the preparative possibilities if the ratio between the rate constants k_p and k_N is sufficiently high.

References

Balavoine, G., Moradpour, A., and Kagan, H. B. (1974). *J. Am. Chem. Soc.* **96**, 5152.

Brandt, J., Jochum, C., Ugi, I., and Jochum, P. (1977). *Tetrahedron* **33**, 1353.

Bredig, G., and Fajans, K. (1908). *Ber. Dtsch. Chem. Ges.* **41**, 752.

Brugidou, J., Christol, H., and Sales, R. (1974). *Bull. Soc. Chim. Fr.*, p. 2027.

Christol, H., Duval, O., and Solladie, G. (1968). *Bull. Soc. Chim. Fr.*, p. 4151.

Conan, J. Y., Nataf, A., Guinot, G., and Lamaty, G. (1974). *Bull. Soc. Chim. Fr.*, pp. 1400, 1405.

Fajans, K. (1910). *Z. Phys. Chem.* **73**, 25.

Horeau, A. (1961). *Tetrahedron Lett.*, p. 605.

Horeau, A. (1964a). *J. Am. Chem. Soc.* **86**, 3171.

Horeau, A. (1964b). *Bull. Soc. Chim. Fr.*, p. 2673.

Horeau, A. (1969). *Tetrahedron Lett.*, p. 3121.

Horeau, A. (1975). *Tetrahedron* **31**, 1307.

Horeau, A. (1977). *In* "Stereochemistry, Fundamentals and Methods" (H. B. Kagan, ed.). Thieme, Stuttgart.

Horeau, A., and Nouaille, A. (1966). *Tetrahedron Lett.*, p. 3953.

Horeau, A., Nouaille, A., and Mislow, K. (1965). *J. Am. Chem. Soc.* **87**, 4958.

Horeau, A., Guetté, J. P., and Weidmann, R. (1966). *Bull. Soc. Chim. Fr.*, p. 3513.

Jacques, J., Collet, A., and Wilen, S. H. (1981). "Enantiomers, Racemates and Resolutions." Wiley, New York.

Kagan, H. B., Balavoine, G., and Moradpour, A. (1974). *J. Mol. Evol.* **4**, 41.

King, J. F., and Sim, S. K. (1973). *J. Am. Chem. Soc.* **95**, 4448.

King. J. F., Sim, S. K., and Li, S. K. L. (1973). *Can. J. Chem.* **51**, 3914.

Kuhn, W., and Knopf, E. (1930). *Z. Phys. Chem. Abt. B* **7**, 292.

LeBel, J. A. (1874). *Bull. Soc. Chim. Fr.* **22**, 337.

Marckwald, W., and McKenzie, A. (1899). *Ber. Dtsch. Chem. Ges.* **32**, 2131.

Martin V. S., Woodward, S. S., Katsuki, T., Yamada, Y., Ikeda, M., and Sharpless, K. B. (1981). *J. Am. Chem. Soc.* **103**, 6237.

Meurling, L., and Bergson, G. (1974). *Chem. Scr.* **6**, 104.

Meurling, L., Bergson, G., and Obenius, U. (1976). *Chem. Scr.* **9**, 9.

Mislow, K. (1966). "Introduction to Stereochemistry," p. 128. Benjamin, New York.

Newman, P., Rutkin, P., and Mislow, K. (1958). *J. Am. Chem. Soc.* **80**, 465.

Schoofs, A., and Horeau, A. (1977a). *Tetrahedron* **33**, 245.

Schoofs, A., and Horeau, A. (1977b). *Tetrahedron Lett.*, p. 3259.

4

Isotope-Dilution Techniques

Kenneth K. Andersen
Diana M. Gash
John D. Robertson
Department of Chemistry
University of New Hampshire
Durham, New Hampshire

I. Introduction

The method of isotope dilution can be used to determine the enantiomeric excess (ee)[1] of a sample. This technique proceeds as described on the following page:

[1]There has been some debate (Valentine and Scott, 1978) about how best to express the ratios of enantiomers in mixtures. Because in this chapter we are attempting to stress the applicability of this technique to a variety of detection methods and to the use of isotopic labels other than radioactive ones, optical purity P and the expression for excess enantiomer E used by Berson and co-workers are considered inappropriate. For this reason the expression *enantiomeric excess* (ee) has been used wherever possible:

$$ee = E/(E + R) = E/y$$

or

$$\% \ ee = [E/(E + R)] \times 100 = 100E/y.$$

For definitions of E, R, and y, refer to Table I and Section III.

1. An isotopically labeled sample of known enantiomeric purity, usually a racemic modification (Eliel, 1962), is mixed with a test sample of unknown enantiomeric purity.
2. A sample is isolated from the mixture, and its isotope content is analyzed.
3. The enantiomeric purity of the test sample is calculated from the dilution of the isotopic label. By including optical rotation measurements, the absolute rotations of the enantiomers can be calculated.

II. History

Isotope-dilution analysis was first described by Hevesy and Hobbie (1932) as a method of analyzing the lead content of ore. Two years later, in a delightful letter to *Nature,* Hevesy and Hofer (1934) reported the results of a study in which deuterium water was used as a label to determine the total water content of the human body and also to measure the residence time of water in the body.

It was not until Rittenberg and Foster (1940) described the use of isotope dilution for the analysis of complex mixtures of organic compounds that the technique was shown to have great analytical utility (Raban and Mislow, 1967). These workers also recognized the potential of the method for determining the relative D or L content of isolated amino acids and, in a subsequent paper (Rittenberg *et al.*, 1940), published results of the determination of the D- and L-glutamic acid content of malignant tumor tissue (see also Ussing, 1939).

TABLE I

Symbols Used in Equations for Determination of Enantiomeric Purity

x	Known amount of labeled sample
C_0	Isotope content of x
y	Unknown amount of substance to be determined
y_+	Amount of $(+)$-enantiomer in y (Fig. 3)
y_-	Amount of $(-)$-enantiomer in y (Fig. 3)
C	Isotope content of isolated sample
$C_\pm$	Isotope content of isolated racemic modification (Fig. 4)
C_+	Isotope content due to $(+)$-enantiomer in sample before isolation (Figs. 4 and 5)
C_-	Isotope content due to $(-)$-enantiomer in sample before isolation (Figs. 4 and 5)
C_{+2}	Isotope content due to $(-)$-enantiomer in isolated sample (Fig. 5)
C_{-2}	Isotope content due to $(+)$-enantiomer in isolated sample (Fig. 5)
R	Amount of y that is a racemic modification (Fig. 5)
E	Amount of y that is an excess of one enantiomer (Fig. 5)
R_2	Amount of isolated sample (Fig. 5) that is a racemic modification
E_2	Amount of isolated sample (Fig. 5) that is an excess of one enantiomer
Z	Mass of isolated sample (Fig. 5)

In 1959 Berson and Ben-Efraim published the first account of the application of isotope dilution to the determination of enantiomeric purity. They developed equations for determining the enantiomeric purity of a mixture and, when used in conjunction with measured optical rotations, for calculating the enantiomeric purity of isolated enantiomers. In later work (Berson and Suzuki, 1959; Berson and Willner, 1964), isotope dilution was again used to check the enantiomeric purity of chiral substrates.

During the 1960s other workers (Cope et al., 1966; Gerlach, 1966; Goering and Doi, 1960; Goering and Towns, 1963; Goering et al., 1963, 1964; Kemp et al., 1970) employed the technique in conjunction with chiral compounds, but after this period there were very few studies (Baggiolini et al., 1970; Falk and Lerner, 1971; Farina and Audisio, 1970; Knabe and Gradmann, 1973) that used isotope dilution for finding the enantiomeric purity of optically active mixtures. More recent papers published in the field are those of German groups (Spielmann and de Meijere, 1976; Von Stephan et al., 1976).

III. Theory

In this section the equations used in the determination of enantiomeric purity by the method of isotope dilution are derived (Table I). Consider first the situation in which it is necessary to determine the amount of compound A in a mixture. Suppose that it is possible to isolate a pure sample of A from the mixture but impossible to do this quantitatively. If a known quantity of an isotopically labeled sample of pure A (let this be designated A*, where the asterisk refers to the label) is added to the mixture, then A* will be diluted by the unlabeled A in the mixture. A pure sample (m) isolated from the diluted mixture will contain both A and A*. The ratio A/A* in the isolated sample will be identical to the ratio A/A* in the diluted mixture if isotope effects on the isolation are negligible, as they usually are. Since the amount of A* is known, the amount of A can be

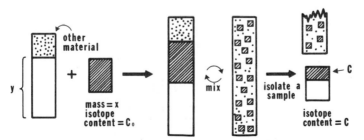

Fig. 1. Determination of the quantity of one component in a mixture.

calculated (Marshall, 1976). The label has in many examples been a radioisotope, but any isotope that can be determined quantitatively is suitable.

The procedure can be represented diagrammatically (Fig. 1). A known amount x (for convenience this will be expressed as a mass, but it could be a volume or any other convenient measure; see Marshall, 1976) of A* containing a known concentration C_0 of isotopic label is added to a mixture containing an unknown amount y of A. A pure sample of A + A* is isolated from this mixture. If the specific isotope concentration C of the isolated sample is determined, y can be calculated:

$$C/C_0 = x/(x + y) \tag{1}$$

or

$$y = x \left(\frac{C_0}{C} - 1\right). \tag{2}$$

The extension of this to mixtures of enantiomers requires that a racemic modification, a mixture enriched in one enantiomer, or a pure enantiomer can be isolated. The labeled tracer is usually synthesized as a racemic modification, symbolized by $(\pm)^*$.

First, consider adding a known quantity x of $(\pm)^*$ with an isotope content C_0 to an unlabeled racemic modification $(\pm)$ (Fig. 2). For clarity in the diagrams the labeled material is shaded. The equation derived is identical to that obtained for the first dilution experiment [Eq. (1)].

Consider now an enantiomeric mixture of $(+)$ and $(-)$ enriched in one enantiomer. It is convenient to split up the mixture and consider each enantiomer separately. When $(\pm)^*$ is added to the mixture, the $(+)^*$ is diluted by the $(+)$ and the $(-)^*$ is diluted by the $(-)$.

The mass y of $(+)$ and $(-)$ can be considered to be composed of a racemic modification R and an excess of enantiomer E [Fig. 3, where y_+ and y_- are the masses of $(+)$ and $(-)$, respectively]. If the sample isolated is a racemic modifi-

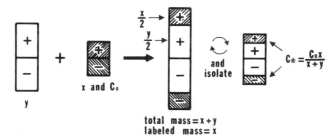

Fig. 2. Addition of a known quantity of labeled racemic modification to an unlabeled racemic modification.

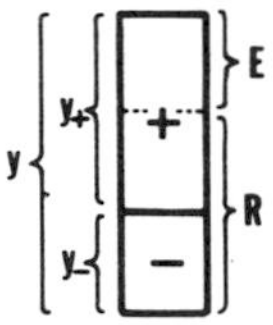

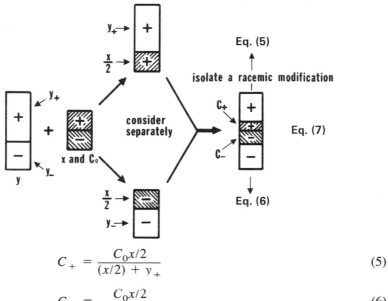

$$y_+ = y - \frac{R}{2} = E + \frac{R}{2} \tag{3}$$

$$y_- = R/2 \tag{4}$$

Fig. 3.   Relationships between E, R, y, y_+, and y_-.

$$C_+ = \frac{C_0 x/2}{(x/2) + y_+} \tag{5}$$

$$C_- = \frac{C_0 x/2}{(x/2) + y_-} \tag{6}$$

$$C_\pm = \tfrac{1}{2}\left(\frac{C_0 x}{x + 2y_+}\right) + \tfrac{1}{2}\left(\frac{C_0 x}{x + 2y_-}\right) \tag{7}$$

Fig. 4.   Determination of the enantiomeric excess (ee) of a sample. In Eq. (7), because a racemic modification is isolated, the weighting factor is $\tfrac{1}{2}$.

cation, Eq. (8) may be derived. For the purposes of this derivation the (+)-enantiomer is considered to be in excess (Fig. 4). Substituting Eqs. (3) and (4) into Eq. (7) gives

$$C_\pm = \frac{C_0 x}{2(x + R + 2E)} + \frac{C_0 x}{2(x + R)} . \tag{8}$$

If Eq. (8) is written in terms of ee, it becomes

$$C_\pm = \frac{C_0 x}{2[x + y(1 + ee)]} + \frac{C_0 x}{2[x + y(1 - ee)]} . \tag{9}$$

There is the possibility of preparing a sample of pure isotopically labeled enantiomer from, for example, a natural product. In this case, assuming as above that the (+)-enantiomer is prepared, Eq. (9) becomes

$$C_{\pm} - \frac{C_0 x}{2[x + (y/2)(1 + ee)]} .$$
(10)

Equation (9) can be rearranged thus:

$$ee = \frac{1}{y}\left[(x + y)^2 - \frac{C_0 x}{C_{\pm}}(x + y) \right]^{1/2}$$
(11)

or

$$(ee)^2 = \frac{C_{\pm}(x + y)^2 \quad C_0 x(x + y)}{y^2 C_{\pm}} .$$
(12)

Consider now the situation in which the isolated material is not a racemic modification (a labeled racemic modification is still used as the tracer; Fig. 5):

$$C_+ = \frac{C_0 x/2}{x/2 + R/2 + E} = \frac{C_0 x}{x + R + 2E} ,$$
(13)

$$C_- = \frac{C_0 x/2}{x/2 + R/2} = \frac{C_0 x}{x + R} .$$
(14)

Equations (13) and (14) expressed in terms of ee become

$$C_+ = \frac{C_0 x}{x + y(1 + ee)}$$
(15)

and

$$C_- = \frac{C_0 x}{x + y(1 - ee)} .$$
(16)

An approach similar to that used in deriving Eqs. (5) and (6) is used when one is considering the isolated sample; that is, it is considered to be racemic modifi-

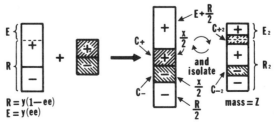

Fig. 5. Extension of the procedure shown in Fig. 4 to a situation in which the isolated material is not a racemic modification.

cation, and an excess of one enantiomer [again, for convenience, the (+)-enantiomer] is chosen (Fig. 5). Hence,

$$R_2 = Z - E_2. \tag{17}$$

Also,

$$C_{+2} = C_+ \frac{R_2/2 + E_2}{Z} \tag{18}$$

and

$$C_{-2} = C_- \frac{R_2}{2Z} . \tag{19}$$

Substituting Eq. (17) into Eqs. (18) and (19) gives

$$C_{+2} = \frac{C_+}{2} \left(1 + \frac{E_2}{Z} \right) \tag{20}$$

and

$$C_{-2} = \frac{C_-}{2} \left(1 - \frac{E_2}{Z} \right) . \tag{21}$$

Therefore,

$$C = C_{+2} + C_{-2} = \frac{C_+}{2} \left(1 + \frac{E_2}{Z} \right) + \frac{C_-}{2} \left(1 - \frac{E_2}{Z} \right) . \tag{22}$$

However,

$$(ee)_2 = E_2/2 \tag{23}$$

so, substituting Eq. (23) into Eq. (22),

$$C = \frac{C_+}{2} [1 + (ee)_2] + \frac{C_-}{2} [1 - (ee)_2] . \tag{24}$$

Finally, Eqs. (15) and (16) can be substituted into Eq. (24), and the specific isotope concentration of the isolated material related to the ee of the original mixture is

$$C = C_0 x \frac{(x + y) - y(ee)(ee)_2}{(x + y)^2 - y^2(ee)^2} . \tag{25}$$

This expression is often used to determine the rotations $[\alpha_A]$ of the original enantiomers by measuring rotations of the unresolved mixtures before, $[\alpha_1]$, and after, $[\alpha_2]$, the isotope-dilution experiment. Because the original mixture and the reisolated material contain only enantiomers, we can write

$$ee = [\alpha_1]/[\alpha_A] \tag{26}$$

and

$$(ee)_2 = [\alpha_2]/[\alpha_A].$$ (27)

Substituting this in Eq. (25) and making $[\alpha_A]$ the subject of the equation gives

$$[\alpha_A]^2 = \frac{Cy^2[\alpha_1]^2 - C_0xy[\alpha_1][\alpha_2]}{C(x + y)^2 - C_0x(x + y)} \cdot$$ (28)

If the reisolated sample is racemic, $[\alpha_2] = 0$. Hence,

$$[\alpha_A]^2 = \frac{Cy^2[\alpha_1]^2}{C(x + y)^2 - C_0x(x + y)} \cdot$$ (29)

This is equivalent to Eq. (12).

Goering and Doi (1960) and also Berson and Willner (1964) used a refinement of Eq. (28) in which a sample was resolved to give what was thought to be a pure enantiomer, $[\alpha_1]$ was measured, and then labeled racemic modification was added and the sample re-resolved until the measured $[\alpha_2]$ was equal to $[\alpha_1]$. Thus,

$$[\alpha_A]^2 = \frac{(Cy^2 - C_0xy)[\alpha_1]^2}{C(x + y)^2 - C_0x(x + y)} \cdot$$ (30)

Note that, as long as $[\alpha_1] = [\alpha_2]$, this would work for any degree of enantiomeric purity.

Gerlach (1966) expressed Eq. (29) as

$$[\alpha_A]^2 = \frac{Q[\alpha_1][\alpha_3] - [\alpha_3]^2}{Q - 1},$$ (31)

where

$$[\alpha_3] = [\alpha_1]y/(x + y)$$ (32)

and

$$Q = C_3/C = C_0x/C(x + y).$$ (33)

Here, $[\alpha_3]$ is the rotation of the test sample after the addition of labeled racemic modification but before reresolution, and Q is the isotopic content of the sample after the addition of racemic modification but before reresolution (C_3) divided by the isotopic content after final resolution.

IV. Errors

Errors of 1 to 3% in ee and $[\alpha]$ are reported in the papers cited, regardless of the molecular system or analytical technique employed. Attention must be paid to the following points in order to minimize error:

1. The diluent and recovered sample must normally be chemically pure.
2. The isotopic concentration of the diluent and recovered sample must be accurately known. The most significant errors occurred here for the three possible determination methods, namely, scintillation counting of radioisotopes (^{14}C or ^{36}Cl), ^{1}H-NMR spectroscopy, and mass spectrometry (MS).
3. If the isotope used has a high natural abundance, the proportion of isotopic label incorporated into the diluent must be higher than the proportion used if the natural abundance of the isotope is low (Rittenberg and Foster, 1940).
4. The ratio of diluent to unknown or vice versa must be such that the difference in isotopic concentration between the diluent and recovered sample is not too small.

Of the three detection methods just mentioned, MS and especially GC–MS (Jenden, et al., 1973; Campbell, 1974; Pickup and McPherson, 1976; Colby and McCaman 1979; Schramm et al., 1979) have been used extensively for the analysis of biological systems and for inorganic trace metal analysis via isotope-dilution methods (Tölgyessy and Varga, 1972). The increasing convenience and sensitivity of MS and GC–MS analysis together with the ready availability of stable isotopes that pose no storage and disposal problems appear to make this the method of choice for ee determinations by isotope-dilution techniques.

V. Examples

In the first published example of the use of isotope dilution as an aid in the determination of enantiomeric purity, Berson and Ben-Efraim (1959) established the composition of an optically active ($[\alpha]_D^{23}$ measured) mixture of enantiomers of endo-norbornanecarboxylic acid (1) and then calculated the absolute rotation of (+)-endo-norbornanecarboxylic acid. Comparison of the calculated $[\alpha]_D^{25}$ with $[\alpha]_D^{25}$ obtained by polarimetry allowed the optical purity of (+)-endo-norbornanecarboxylic acid to be determined. In this work ^{14}C was used as an isotopic label. The activity of the diluted and undiluted samples was measured using scintillation counting.

COOH

1

A sample of racemic *endo*[2,3-^{14}C]norbornanecarboxylic acid was synthesized (for details see Berson and Ben-Efraim, 1959) and found to have an activity of 2965 $\pm$ 12 counts per minute (C_0). To 3.8104 g (*x*) of this labeled material was added 3.7325 g (*y*) of optically active *endo*-norbornanecarboxylic acid, $[\alpha]_D^{25}$ +16.7°. Two recrystallizations from acetonitrile gave labeled racemate with an activity of 1615 $\pm$ 12 counts per minute ($C_\pm$). Substitution of the appropriate numbers into Eq. (12) (Section III) gave an ee value of 51.8 to 54.4% for the test sample of $[\alpha]_D^{25}$ +16.7°. This gave $[\alpha]_D^{25}$ +30.6–32.1° for enantiomerically pure (+)-*endo*-norbornanecarboxylic acid. Because $[\alpha]_D^{25}$ +30.1° was obtained by polarimetry, the enantiomeric purity of a sample of (+)-*endo*-norbornanecarboxylic acid prepared by systematic resolution was at least 94%.

In another example from the literature, Cope *et al.* (1966) checked the enantiomeric purity of the (−)-*N*-*n*-butyl-*N*-isobutyl-*N*-methylcyclooctylammonium cation (**2**) by isotope dilution. Nitrogen-15 was used as an isotopic label, and its

2

dilution was determined by MS. Initially, the isotope-dilution experiment was attempted using the perchlorate salt of **2**. All efforts to crystallize racemic perchlorate from a mixture of racemic ^{15}N-labeled perchlorate (2.0007 g, *x*) and (−)-perchlorate (3.9983 g, *y*) failed, however. Therefore, the perchlorate mixture was converted first to the corresponding hydroxide by ion-exchange chromatography and then to the chloride by neutralization of the hydroxide by dilute hydrochloric acid. Treatment of the chloride with sodium tetraphenylboron gave a mixture of racemic and (−)-tetraphenylboron salts, from which the racemate could be crystallized. Analysis by MS was carried out on the recovered racemate. It was found that the mass spectrum of *N*-2-butyl-*N*-isobutyl-*N*-methylcyclooctylammonium tetraphenylboron shows no molecular ion (*m/z* 574) and has a base peak at *m/z* 100; this could be attributed to an ion (**4**) arising from the cleavage of *N*-isobutyl-*N*-methyl-*n*-butylamine (**3**). Comparison of the region

3   **4**

from m/z 99 to m/z 102 in this spectrum with the same region in the spectrum of pure N-isobutyl-N-methyl-n-butylamine showed that the two regions were identical. In both cases the peak height at m/z 101 was found to be 7.1% of the peak height at m/z 100 and therefore due only to the natural isotopic abundance (the calculated value is 7.09%) (Silverstein et al., 1981). To determine the atom % excess of ^{15}N (C_0) for the undiluted [^{15}N]tetraphenylboron salt and the atom % excess of ^{15}N (C) for the diluted [^{15}N]tetraphenylboron salt, the mass spectra of both and also the mass spectrum of the corresponding [^{14}N]tetraphenylboron salt were measured under identical conditions, scanning the region of the spectrum from m/z 99 to m/z 102 thirty times for each sample. Data are listed in Table II. For the undiluted [^{15}N]tetraphenylboron salt, the atom % excess of ^{15}N (C_0) was calculated as follows:

Total N $= 100 + 70.2 - $ (contribution due to ^{13}C) $+$ (natural abundance of ^{15}N)

$\qquad = 100 + 70.2 - 7.1 + 0.37$

$\qquad = 100 + 63.47$

$\qquad = 163.47;$

$$\% \text{ excess } {}^{15}\text{N} = \frac{63.47}{163.47} \times 100$$

$$= 38.8\%.$$

A value of 23.1 ± 0.2% can be similarly calculated for the atom % excess of ^{15}N (C) for the diluted ^{15}N-labeled tetraphenylboron salt.

Substitution of numerical values for C, C_0, x, and y in Eq. (12) (Section III) gives an ee value of 99.6 ± 0.8% for $(-)$-N-n-butyl-N-isobutyl-N-methylcyclooctylammonium perchlorate.

Deuterium was used as an isotopic label in work carried out by Reich-Rohrwig and Schlögl (1968). In addition to the isotope-dilution method, ^{1}H-NMR spectroscopy was used as a second method for determining the enantiomeric purity of methylferrocene-α-carboxylic acid (5). The results obtained from the two methods were compared and found to be similar.

TABLE II
Mass Spectra of the Tetraphenylboron Salts

Salt	m/z 100 (%)	m/z 101 (%)	m/z 102 (%)
[^{14}N]Tetraphenylboron	100	7.1 ± 0.07	—
[^{15}N]Tetraphenylboron			
Undiluted	100	70.2 ± 0.6	4.6 ± 0.1
Diluted	100	36.8 ± 0.3	2.5 ± 0.06

$$\text{Fe} \overset{\text{—COOH}}{\underset{\text{CH}_3}{}}$$

5

An example of the use of isotope dilution for something other than the determination of enantiomeric purity is found in a paper by Beckwith and Hager (1961). They wanted to prove that naturally occurring caldariomycin was one-half of the synthetic pair of enantiomers **6a** and **6b**. The following radioisotope-

6a   **6b**

dilution experiment was carried out. ^{36}Cl-Labeled natural caldariomycin (40 mg, 860 counts per minute per milligram) was mixed with synthetic DL-caldariomycin (40 mg) and repeatedly crystallized from chloroform until caldariomycin of melting point identical to that of the natural product was obtained. The theoretical specific radioactivity, assuming a dilution of ^{36}Cl-labeled caldariomycin with one-half its weight of unlabeled material, would be 66.6% of its original value (Fig. 6). Experimentally, a value of 572 counts per minute per milligram was obtained for the recrystallized sample. This demonstrated that natural caldariomycin constituted half of the synthetic racemate.

Kemp *et al.* (1970) used the principle of isotope dilution in preparing an isotopically labeled chiral substance containing negligible label in any contaminating enantiomer. This was accomplished by successive dilution of a labeled enantiomer of high enantiomeric purity with a small quantity of unlabeled racemate. If the labeled enantiomer was of 99.5% enantiomeric purity, then dilution of x_1 grams of this with $x_1/10$ grams of unlabeled racemate, followed by

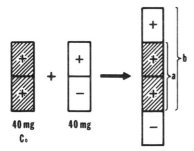

40 mg
C_0   **40 mg**

Fig. 6. Isotope-dilution experiment using labeled caldariomycin. $a/b = \frac{2}{3}$; therefore, $C = \frac{2}{3}C_0$.

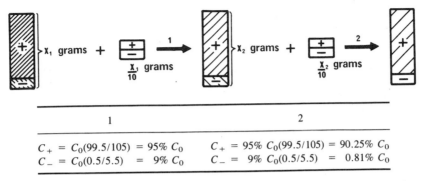

1	2

$C_+ = C_0(99.5/105) = 95\% \ C_0$ $C_+ = 95\% \ C_0(99.5/105) = 90.25\% \ C_0$

$C_- = C_0(0.5/5.5) \quad = 9\% \ C_0$ $C_- = 9\% \ C_0(0.5/5.5) \quad = 0.81\% \ C_0$

Fig. 7. Preparation of an isotopically labeled enantiomer containing negligible label in any contaminating enantiomer.

recovery of labeled enantiomer, would result in dilution of the label to 99.5/105 of its original value (Fig. 7). Any contaminating enantiomer in the original sample, however, would now possess 0.5/5.5 of its original label. Since this process could be repeated, it was possible to attain labeled desired enantiomer: labeled contaminating enantiomer ratios of 10^6 to 10^7.

A labeled sample prepared by this method could be used as a probe to examine the stereospecificity of a given procedure. For example, Kemp and co-workers (1970) used labeled amino acid derivatives to test the stereospecificity of some commonly used amino acid coupling methods.

The preceding examples illustrate some of the applications of the isotope-dilution technique. Although it has most commonly been used as an aid in the determination of enantiomeric purity, the work of Beckwith and Hager (1961) and Kemp et al. (1970), briefly summarized above, demonstrates that isotope dilution can be used to solve other problems.

References

Baggiolini, E., Hamlow, H. P., and Schaffner, K. (1970). *J. Am. Chem. Soc.* **92**, 4906.
Beckwith, J. R., and Hager, L. P. (1961). *J. Org. Chem.* **26**, 5206.
Berson, J. A., and Ben-Efraim, D. A. (1959). *J. Am. Chem. Soc.* **81**, 4083.
Berson, J. A., and Suzuki, S. (1959). *J. Am. Chem. Soc.* **81**, 4088.
Berson, J. A., and Willner, D. (1964). *J. Am. Chem. Soc.* **86**, 610.
Campbell, I. M. (1974). *Bioorg. Chem.* **3**, 386.
Colby, B. N., and McCaman, M. W. (1979). *Biol. Mass Spectrosc.* **6**, 225.
Cope, A. C., Funke, W. R., and Jones, F. N. (1966). *J. Am. Chem. Soc.* **88**, 4693.
Eliel, E. L. (1962). "Stereochemistry of Carbon Compounds," p. 31. McGraw-Hill, New York.
Falk, H., and Lehner, H. (1971). *Tetrahedron* **27**, 2279.

Farina, M., and Audisio, G. (1970). *Tetrahedron* **26**, 1839.
Gerlach, H. (1966). *Helv. Chim. Acta* **49**, 2481.
Goering, H. L., and Doi, J. T. (1960). *J. Am. Chem. Soc.* **82**, 5850.
Goering, H. L., and Towns, D. L. (1963). *J. Am. Chem. Soc.* **85**, 2295.
Goering, H. L., Pombo, M. M., and McMichael, K. D. (1963). *J. Am. Chem. Soc.* **85**, 965.
Goering, H. L., Doi, J. T., and McMichael, K. D. (1964). *J. Am. Chem. Soc.* **86**, 1951.
Hevesy, G., and Hobbie, R. (1932). *Z. Anal. Chem.* **88**, 1.
Hevesy, G., and Hofer, E. (1934). *Nature (London)* **134**, 879.
Jenden, D. J., Roch, M., and Booth, R. A. (1973). *Anal. Biochem.* **55**, 438.
Kemp, D. S., Wang, S. W., Busby, G., III, and Hugel, G. (1970). *J. Am. Chem. Soc.* **92**, 1043.
Knabe, J., and Gradmann, V. (1973). *Arch. Pharm. (Weinheim)* **306**, 306.
Marshall, R. A. G. (1976). *J. Chem. Educ.* **5**, 320.
Pickup, J. F., and McPherson, K. (1976). *Anal. Chem.* **48**, 1885.
Raban, M., and Mislow, K. (1967). *Top. Stereochem.* **2**, 202.
Reich-Rohrwig, P., and Schlögl, K. (1968). *Monatsh. Chem.* **99**, 1752.
Rittenberg, D., and Foster, G. L. (1940). *J. Biol. Chem.* **133**, 737.
Rittenberg, D., Graff, S., and Foster, G. L. (1940). *J. Biol. Chem.* **133**, 745.
Schramm, W., Louton, T., and Schill, W. (1979). *Fresenius Z. Anal. Chem.* **294**, 107.
Silverstein, R. M., Bassler, G. C., and Morrill, T. C. (1981). "Spectrometric Identification of Organic Compounds" 4th ed., p. 51. Wiley, New York.
Spielmann, W., and deMeijere, A. (1976). *Angew. Chem. Int. Ed. Engl.* **15**, 429.
Tölgyessy, J., and Varga, S. (1972). "Nuclear Analytical Chemistry," Vol. 2, p. 60. Univ. Park Press, Baltimore, Maryland.
Ussing, H. H. (1939). *Nature (London)* **144**, 977.
Valentine, D., and Scott, J. W. (1978). *Synthesis* **5**, 329.
Von Stephan, H. M., Langer, G., and Wiegrebe, W. (1976). *Pharm. Acta Helv.* **51**, 164.

5

Gas Chromatographic Methods

Volker Schurig
Institut für Organische Chemie
Universität Tübingen
Tübingen, Federal Republic of Germany

I. Introduction

A. General Remarks

The enantiomeric excess (ee) achieved during the transformation of a prochiral reactant to a chiral product is a quantitative measure of the success of an asymmetric synthesis. Attempts to mimic stereoselectivities encountered in nature have resulted in the development of highly enantiospecific reactions in the laboratory, and reliable analytical techniques for the precise determination of large enantiomeric excesses (i.e., >95% ee) are becoming increasingly important (Schurig et al., 1980).

Usually, the ee is indirectly obtained by measuring the optical rotation of the product and expressing it as the percentage of the maximum rotation α_{max} of the pure enantiomer, that is,

$$\% \text{ ee} = \frac{[\alpha]}{[\alpha_{max}]} 100. \tag{1}$$

However, the determination of "optical purities" by polarimetry may suffer from serious drawbacks, such as the following:

1. The maximum optical rotation $[\alpha_{max}]$ of the pure enantiomer (absolute optical rotation) must be known with certainty.
2. Relatively large sample sizes are required for polarimetric measurements.
3. The product must exhibit medium to high optical rotatory power, permitting the correct determination of small differences in ee.
4. The chiral product must be isolated and purified without accidental enantiomer enrichment.
5. The accuracy of optical rotations depends on temperature, solvent, and traces of optically active (or inactive) impurities.
6. "Optical purities" may not, a priori, conform to enantiomeric compositions (Horeau, 1969).

Alternative methods of determining enantiomeric compositions include isotopic dilution, kinetic resolution, enzymatic assays, microcalorimetric techniques, as well as NMR spectroscopy in chiral solvents (Raban and Mislow, 1967). A highly attractive method for the analysis of asymmetric reactions is based on the resolution of enantiomeric mixtures by gas chromatography. The comparison of relative peak areas provides a precise measurement of enantiomeric composition (ee). The method, with which one can adequately cope with minute samples, is independent of the magnitude of the optical rotatory power and is not affected by impurities.

The merits of gas chromatography, in general, are speed, sensitivity, and simplicity. The merits of gas chromatography for enantiomer resolution, in particular, are high resolution, precision, and reproducibility. Refined gas chromatographic equipment is available in most modern research facilities. Nonetheless, the prerequisites of the use of gas chromatography as an analytical tool in asymmetric synthesis are substrate volatility, thermal stability, and quantitative resolvability, restricting its general use.

B. Resolution of Enantiomers by Gas Chromatography

The resolution of enantiomeric mixtures by gas chromatography can be performed in two modes: (a) conversion of the enantiomers into diastereomeric derivatives by chemical reaction with an auxiliary, enantiomerically pure, chiral resolving agent and subsequent gas chromatographic separation of the resulting diastereomers on an achiral stationary phase (Guetté and Horeau, 1965; Gil-Av and Nurok, 1974; Halpern, 1977); and (b) direct resolution of the enantiomers on a chiral stationary phase containing an auxiliary resolving agent of high enantiomeric purity (Gil-Av, 1975). These methods are really nothing more than variations of Pasteur's classical approach to enantiomer resolution via crystallization of diastereomeric salts. Whereas in method (a) diastereomers are isolated before chromatographic separation, in method (b) direct resolution is effected via the rapid and reversible diastereomeric interaction between the racemic solute and the optically active stationary phase. Of the two approaches to resolution the direct method (b) has proved to be more reliable and is treated in this chapter exclusively.

Method (a) is restricted to substrates that possess at least one reactive function for quantitative reaction with the resolving agent. Difficulties in the chemical reaction step may involve racemization as well as kinetic resolution resulting from energetically different diastereomeric transition states. Another drawback of method (a) is the possibility of accidental fractionation occurring during derivatization, workup, and chromatographic analysis, leading to a change of the original ratio of enantiomers in the sample. Finally, systematic errors may arise from the incomplete enantiomeric purity of the resolving agent, which will affect the accuracy of the % ee determination for highly enriched mixtures.

The combination of methods (a) and (b), that is, the resolution of the preformed diastereomers into enantiomeric pairs (RR'/SS' and RS'/SR') on a chiral stationary phase, may clarify ambiguities related to sample racemization versus incomplete enantiomeric purity of the resolving agent.

Method (b) requires an efficient solute–solvent system capable of chiral recognition through molecular association. Because this technique deals with optical isomers, the enantiomeric ratio of the sample will not be altered by chemical, physical, or analytical manipulations before or after resolution. The only source of a systematic error, if any, may be partial diastereoselective decomposition of the racemate on the chiral stationary phase during chromatographic elution. The enantiomeric ratio of the substrate determined by method (b) is independent of the enantiomeric purity of the chiral stationary phase employed. A low enantiomeric purity of the resolving agent, however, results in small separation factors (α).

Method (b) is especially suitable as an analytical tool for asymmetric synthesis when no sample derivatization is required before gas chromatographic analysis. In this case the chiral reaction product may be investigated *in situ,* that is, without isolation and purification using a minute amount of sample (10^{-8} g). Derivatization procedures are required for nonvolatile or highly polar substrates such as α-amino acids. Occasionally, an appropriate derivatization method is employed to improve resolvability by the introduction of suitable functionalities. If derivatization is necessary, in principle a chiral derivatization agent may be selected [method (*a*)]. The inherent advantages of method (*b*), however, call for the use of an achiral reagent. In both cases the absence of racemization accompanying derivatization must be rigorously established.

If, as required by method (*b*), a quantitative enantiomer resolution on a chiral stationary phase is achieved, the inspection of the chromatogram may reveal valuable information (Table I). Relevant for analytical applications to asymmetric synthesis are the peak parameters (c), (d), and (e).

Because any achiral detection device employed in chromatography will respond equally to optical isomers irrespective of their molecular configuration, the comparison of relative peak areas provides an unambiguous measurement of enantiomeric excess via method (*b*) [see Table I, parameter (c)]. Because peak integration can be performed with a high degree of accuracy, precise data can be obtained. This feature is particularly useful in two significant borderline situations involving (*a*) nearly racemic mixtures, that is, by producing evidence for very small ee's, for example, in experiments concerning the amplification of optical activity in nature, and (*b*) highly enantiomerically enriched mixtures, that is, by providing evidence of very small enantiomeric impurities, for example, in

TABLE I
Gas Chromatographic Parameters of Enantiomer Resolution[a]

Parameter	Definition
(a) Peak retention	A thermodynamic measure of the selective association between solute and solvent ($-\Delta G°$)
(b) Peak separation	A thermodynamic measure of chiral recognition between racemic solute and optically active solvent [$\Delta_{S,R}(\Delta G°)$]
(c) Peak ratio	A precise quantitative measure of the enantiomeric composition of the solute (ee)
(d) Peak assignment	A correlation of solute retention and molecular configuration (assignment of absolute configuration)
(e) Peak coalescence	A kinetic measure of solute enantiomerization during resolution ($\Delta G^{\#}$)

[a]From Schurig and Bürkle (1982).

the realm of "total" asymmetric synthesis, efficient kinetic resolution experiments, and enzymatic reactions. For instance, the accurate determination of >95% ee is of importance in contemporary studies of asymmetric synthesis. The assessment of free-energy differences of activation between competing diastereomeric transition states [Eq. (2)] requires accurate analysis. It has been

$$\Delta\Delta G^{\#} = RT \ln (k_s/k_r) = RT \ln ([S]/[R]) \tag{2}$$

$$ee = \frac{[S] - [R]}{[S] + [R]} \tag{3}$$

estimated (Gil-Av, 1975) that as little as 0.1% of an enantiomeric impurity (i.e., 99.8% ee, $\Delta\Delta G_{291}^{\#} = 4.0$ kcal/mol) can be detected by gas chromatography. It can be reasoned that this figure may even be surpassed in carefully designed gas chromatographic experiments by the use of suitable digital electronic integrating devices.

The assignment of absolute configuration is another important parameter in asymmetric synthesis. Contrary to NMR spectroscopy in chiral solvents (Sullivan, 1978), in which chemical shift nonequivalence of enantiotopic nuclei may arise from two independent individual contributions, that is, by different molecular geometries and by different stabilities of the diastereomeric substrate–solvent adducts, resolution of enantiomers by chromatography depends solely on the disparity between the stability constants of the diastereomeric solute–solvent intermediates formed during the elution event. The gas chromatographic correlation of molecular configuration and order of peak emergence would therefore appear to be more straightforward [see Table I, parameter (d)]. Although in many instances satisfactory agreement between molecular configuration and order of elution from optically active stationary phases has been observed for members of certain series of compounds, notable exceptions that demonstrate the limitations of this approach have also been detected (Schurig and Bürkle, 1982).

The configurational integrity of the enantiomers during the gas chromatographic process of resolution is a *sine qua non* for the precise determination of enantiomeric composition (ee). Enantiomerization of the solute, that is, inversion of configuration of the antipodes during elution, will produce transition phenomena in the gas chromatogram [see Table I, parameter (e)], which can be recognized by the appearance of a plateau between the terminal peaks of the enantiomeric fractions (Schurig et al., 1979; Schurig and Bürkle, 1982).

In the remainder of this chapter methods for the gas chromatographic resolution of enantiomers on chiral stationary phases are described, and applications of the technique are discussed.

II. Direct Resolution of Derivatized Enantiomers on Chiral Stationary Phases by Gas Chromatography

A. Methods

Early claims in the literature (Karagounis and Lippold, 1959; Karagounis and Lemperle, 1962) regarding the marginal resolution of racemic mixtures on optically active stationary phases by gas chromatography have been disputed (Coleman et al., 1966). The first successful resolution of racemic N-TFA-amino acid esters (TFA, trifluoroacetyl) on glass capillary columns coated with N-TFA-L-isoleucine lauryl ester (1; Scheme 1) was performed by Gil-Av et al. (1966, 1967). Although the separation factors α of enantiomer resolution were small and column efficiency was low, the great potential of this fundamental approach was apparent. The many efforts to improve resolution have been reviewed up to 1974 (Lochmüller and Souter, 1975). Superior chiral stationary phases, with additional NH groups participating in chiral recognition via hydrogen bonds, have been developed [Scheme 1, dipeptide phase (2), carbonylbis(amino acid) phase (3), and diamide phase (4)] (Feibush and Gil-Av, 1967, 1970; Gil-Av and Feibush, 1967; Nakaparksin et al., 1970; Parr et al., 1970; Feibush et al., 1970; Feibush, 1971; Beitler and Feibush, 1976). The diamide phase (4), which is commercially available (Supelco, Inc., Bellafonte, Pennsylvania, SP-300 coated on Supelcoport; see Bulletin 765), appears to be the most promising resolving agent for N-TFA-amino acid esters with short packed columns. Improved ther-

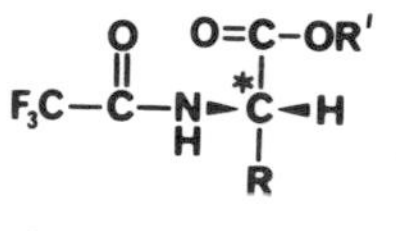

1 R = sec-butyl*;
   R′ = dodecyl

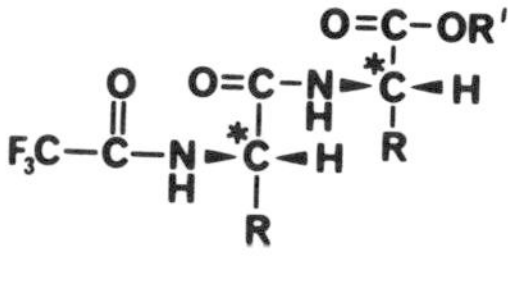

2 R = isopropyl;
   R′ = cyclohexyl

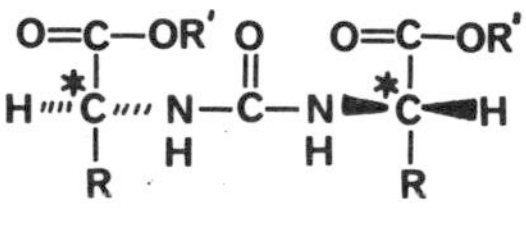

3 R = R′ = isopropyl

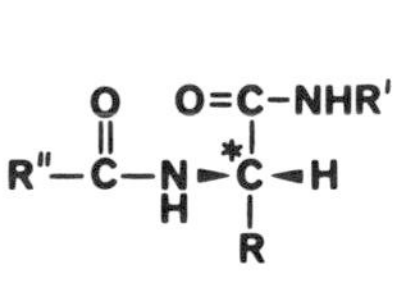

4 R = isopropyl;
   R′ = tert-butyl;
   R″ = undecyl($C_{11}H_{23}$), ($C_{21}H_{43}$)

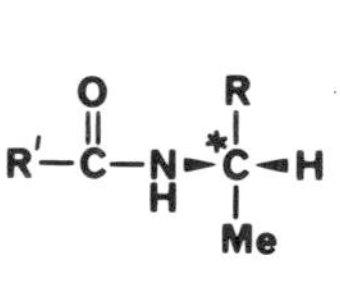

5 R = 1-naphthyl
   R′ = undecyl

Scheme 1. Basic structural types of optically active stationary phases for gas chromatographic enantiomer resolution via hydrogen bonding.

mal stability (up to 190°C) has been achieved with carefully purified preparations of the diamide phase N-docasanoyl-L-valine-tert-butylamide (Charles et al., 1975) and related diamide phases (Charles and Gil-Av, 1980; Chang et al., 1980). The high resolution factors reported for the diamide phases (4) may permit small-scale preparative separations to be performed.

In addition to α-amino acids, β- and γ-amino acids (Feibush et al., 1972) and amines (Feibush and Gil-Av, 1967) have also been resolved. A novel type of chiral stationary phase, N-lauroyl-(S)-α-(1-naphthyl)ethylamine (5, Scheme 1), permitted the gas chromatographic resolution of aromatic and aliphatic N-TFA-amines as well as α-methyl- and α-phenylcarboxylic acid amines (Weinstein et al., 1976). α-Methyl-α-amino acid derivatives have also been resolved (Chang et al., 1982).

Perfluoroacyl derivatives of amino acid esters elute from capillary columns over a broad temperature range owing to differences in volatility, polarity, and hydrogen-bonding interaction with the chiral stationary phase. Although it usually suffices for analytical purposes in asymmetric synthesis to establish optimized conditions for the quantitative resolution of one solute, certain biochemical applications necessitate the simultaneous resolution of all naturally occurring amino acids in one analytical run. To this end, the low thermal stability and appreciable volatility of some of the low-molecular-weight stationary phases restricts their general use. Frank et al. (1977) therefore developed thermally stable and nonvolatile chiral stationary phases by coupling the versatile diamide phase (4) of Feibush (1971) via the amino function to a statistical copolymer of dimethylsiloxane and (2-carboxypropyl)methylsiloxane of high viscosity and appropriate average molecular weight. The fluid polymeric phase (6), referred to as Chirasil-Val (see Scheme 2), shows very high efficiency for enantiomer resolution up to 240°C when coated on pretreated glass capillary columns (Nicholson et al., 1979). Several classes of compounds were baseline-resolved on 6, for example, α-hydroxy acids such as O-PFP-lactic acid cyclohexylamide (PFP, pentafluoropropanoyl) (Frank et al., 1978a), aromatic amino alcohols such as bis-PFP-2-amino-1-phenylethanol and bis-PFP-ephedrine (Frank et al., 1978b), aromatic diols such as bis-PFP-phenylglycol and bis-PFP 2,2' binaphthol, aromatic amines such as PFP-α-phenylethylamine and bis-PFP-2,2'-diamino-1,1'-binaphthyl, as well as an underivatized diketone 3,4-diphenyl-2,5-hexanedione (Koppenhoefer, 1980). The resolution of all racemic protein amino acids on a glass capillary column (20 m × 0.25 mm) coated with Chirasil-Val (6) is reproduced in Fig. 1. The highest resolution and thermal stability are observed when the chiral side chains in 6 are separated by an average of seven dimethylsiloxane units (Bayer and Frank, 1980). It should be noted that an additional chiral center is present in the 2-carboxypropyl side group of 6. Its influence on chiral recognition has not been studied.

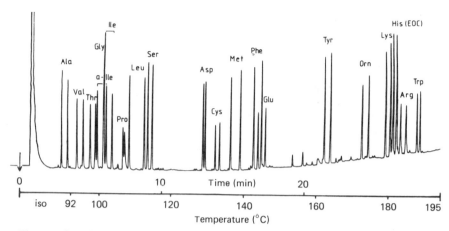

Fig. 1. Gas chromatographic separation of the enantiomers of all protein amino acids (as *N*-PFP-isopropyl esters) on a glass capillary coated with **6** (Chirasil-Val; 20 m × 0.25 mm); temperature progression = 4°C/min. From Bayer and Frank (1980). Copyright 1980 American Chemical Society.

A straightforward approach to a fluid stationary phase of high thermal stability has been devised by Saad *et al.* (1979, 1980). The cyano groups of commercially available poly(cyanopropylmethyl)methylphenylmethylsiloxane (OV-225 or Silar 10C) were hydrolyzed and the carboxy functions were coupled with L-valine-*tert*-butylamide (Feibush, 1971) via the amino function. Aided by an efficient glass capillary technology, various *N*-TFA-amino acid isopropyl esters have been resolved between 60 and 230°C on this phase (**7**, Scheme 2).

Koenig *et al.* (1981a) have coupled *N*-benzyloxycarbonyl-L-valine as well as *N*-hexanoyl-L-valine via the carboxy function to OV-225, in which the cyano groups were reduced with $LiAlH_4$ to the primary amine to give polymer **8** (Scheme 2). Amino acids have been quantitatively resolved on polymer **8** as *N*-TFA-*O*-isopropyl derivatives. A further improvement of resolution was obtained with a fluid polymeric stationary phase prepared by coupling the carboxy functions of hydrolyzed XE-60 polysiloxane to (*S*)-valine-(*S*)- [or (*R*)-] phenylethylamide via the amino group to give **9** (Scheme 2) (Koenig *et al.*, 1981a).

Koenig and Benecke (1981) and Koenig *et al.* (1981a) have also resolved TFA derivatives of aliphatic amines on benzyloxycarbonyl-L-valine coupled to OV-225 (**8**) and on (*S*)-valine-(*R*)-α-phenylethylamide coupled to XE-60 (**9**). (*S*)-Valine-(*S*)-α-phenylethylamide coupled to XE-60 also permitted the resolution of constitutional (i.e., furanose and pyranose) and configurational (i.e., α- and β-epimers) isomers of carbohydrate enantiomers, for example, of *O*-TFA-aldoses and *O*-TFA-methyl glycosides (Koenig *et al.*, 1981b,c).

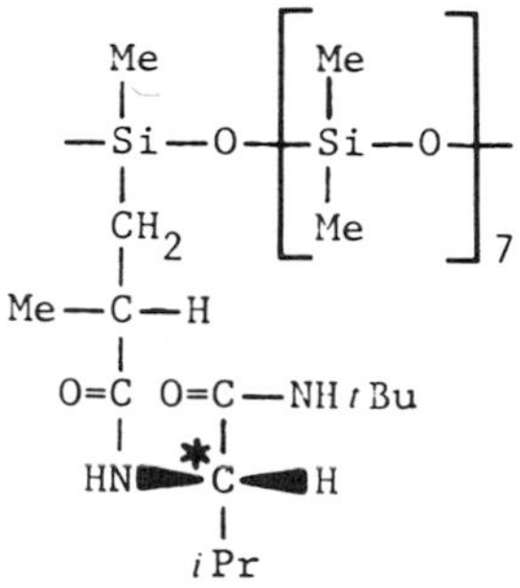

```
        Me          ┌ Me      ┐
        │           │ │       │
      ─Si─O─┼─Si─O─┤
        │           │ │       │
        CH₂         │ Me      │
                    └         ┘₇
    Me─C─H
        │
      O=C  O=C─NH tBu
        │   │
       HN◀─*C─▶H
            │
           iPr
```

6 (Chirasil-Val)

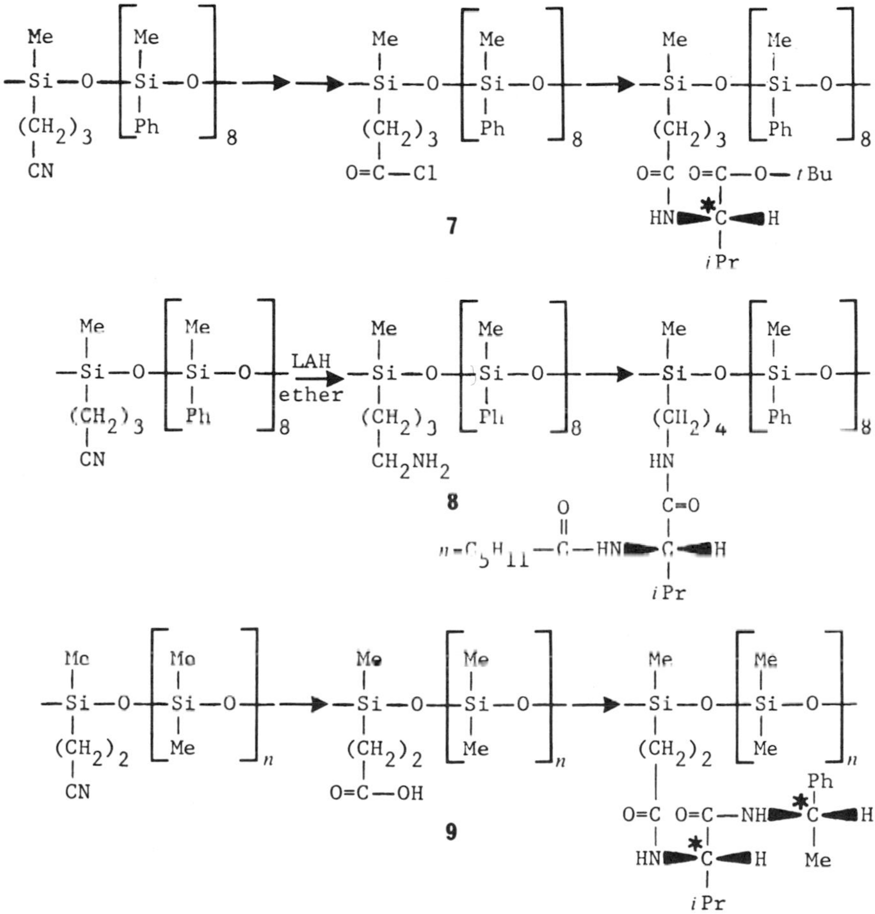

```
  Me     ┌ Me  ┐           Me    ┌ Me  ┐            Me    ┌ Me  ┐
  │      │ │   │           │     │ │   │            │     │ │   │
─Si─O─┼Si─O─┤ ──▶──▶  ─Si─O─┼Si─O─┤ ──▶    ─Si─O─┼Si─O─┤
  │      │ │   │           │     │ │   │            │     │ │   │
(CH₂)₃   │ Ph  │         (CH₂)₃  │ Ph  │          (CH₂)₃  │ Ph  │
  │      └     ┘₈          │     └     ┘₈           │     └     ┘₈
  CN                       O=C─Cl                   O=C  O=C─O─ tBu
                                                     │
                              7                    HN◀─*C─▶H
                                                        │
                                                       iPr
```

```
  Me     ┌ Me  ┐           Me    ┌ Me  ┐            Me    ┌ Me  ┐
  │      │ │   │    LAH     │     │ │   │            │     │ │   │
─Si─O─┼Si─O─┤ ──────▶ ─Si─O─┼Si─O─┤ ──▶     ─Si─O─┼Si─O─┤
  │      │ │   │   ether    │     │ │   │            │     │ │   │
(CH₂)₃   │ Ph  │         (CH₂)₃  │ Ph  │          (CH₂)₄  │ Ph  │
  │      └     ┘₈          │     └     ┘₈           │     └     ┘₈
  CN                       CH₂NH₂                   HN
                                                     │
                              8          O          C=O
                                         ║           │
                              n-C₅H₁₁─C─HN─C◀─▶H
                                                     │
                                                    iPr
```

```
  Me     ┌ Me  ┐           Me    ┌ Me  ┐            Me    ┌ Me  ┐
  │      │ │   │           │     │ │   │            │     │ │   │
─Si─O─┼Si─O─┤ ──▶     ─Si─O─┼Si─O─┤ ──▶     ─Si─O─┼Si─O─┤
  │      │ │   │           │     │ │   │            │     │ │   │
(CH₂)₂   │ Me  │         (CH₂)₂  │ Me  │          (CH₂)₂  │ Me  │
  │      └     ┘ₙ         │     └     ┘ₙ           │     └     ┘ₙ
  CN                       O=C─OH                                 Ph
                                                   O=C  O=C─NH◀─*C─▶H
                              9                     │              │
                                                   HN◀─*C─▶H       Me
                                                        │
                                                       iPr
```

Scheme 2. Structures of polymeric optically active stationary phases for gas chromatographic enantiomer resolution via hydrogen bonding.

V. Schurig

The aliphatic α-hydroxy acid, lactic acid, has been quantitatively resolved on
6 either as its *O*-PFP-cyclohexylamide (Frank *et al.*, 1978b) or as the ethyl ester
after derivatization with *tert*-butyl isocyanate (Hintzer *et al.*, 1982). The introduction of isopropyl isocyanate as a derivatization agent for amines, alcohols,
and hydroxy acids (Koenig *et al.*, 1982a,b) has considerably improved the resolvability of this class of compounds on **9** as a result of additional NH participa-
tion in hydrogen bonding to the diamide stationary phase, which increases chiral
recognition. Thus racemic aliphatic, aromatic, and monoterpene alcohols have
been quantitatively resolved as isopropylurethanes on a 40 m × 0.2 mm glass
capillary column coated with **9**. In Fig. 2 the enantiomer resolution of α-hydroxy
acids (as isopropyl esters and isopropylurethanes), and in Fig. 3 the enantiomer
resolution of *sec*-amines (as isopropylureas), on **9** is reproduced (Koenig *et al.*,
1982a).

Ôi *et al.* (1981a) have reported the resolution of *underivatized* aromatic alcohols such as 1-phenylethanol and 1-phenyl-2,2,2-trifluoroethanol on modified
dipeptide phases of type **2**. It should be noted that the first resolution of an
underivatized aliphatic alcohol, that is, *tert*-butylmethylcarbinol (quantitatively)

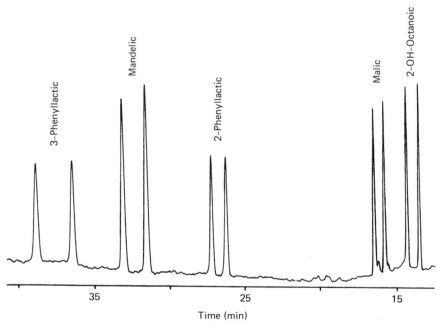

Fig. 2. Gas chromatographic separation of enantiomers of α-hydroxy acids (as isopropyl esters and isopropylurethanes) on a glass capillary coated with **9** (40 m) at 170°C.
From Koenig *et al.* (1982a).

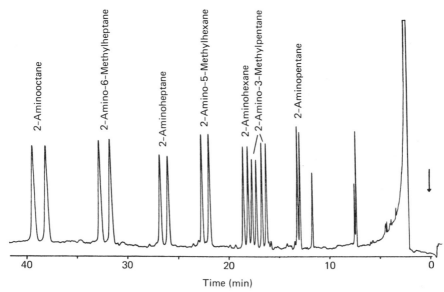

Fig. 3. Gas chromatographic separation of enantiomers of *sec*-amines (as isopropylureas) on a glass capillary coated with **9** (40 m) at 160°C. From Koenig *et al.* (1982a).

or ethylmethylcarbinol (semiquantitatively), has been observed by complexation gas chromatography (Section III,A) on optically active metal chelates (Weber, 1979).

B. Applications

There are very few examples of the use of gas chromatography to determine enantiomeric compositions in asymmetric synthesis. In order to demonstrate the merits of this technique, applications in related areas of chemical research are discussed here.

The first successful resolution of enantiomers by gas chromatography involved amino acids as solutes, and consequently many applications of the technique have dealt with this class of compounds. Thus, the configurations of amino acid components have been determined in biological polymers, biological fluids, extraterrestial material, sediments, and soils (Bayer *et al.*, 1970; Oró *et al.*, 1970; Gil-Av, 1975; Pandey *et al.*, 1977). A geochronometric dating experiment via the racemization of natural amino acids has been carried out on the famous Dead Sea scrolls (E. Gil-Av, personal communication). Small enantiomeric excesses of amino acids in experiments devoted to the amplification of optical activity in abiotic systems have been investigated by gas chromatography (Bonner, 1979). The configurational stability of protein amino acids during peptide hydrolysis and derivatization procedures has been scrutinized by Frank et al (1981) on Chirasil-Val (**6**).

The enantiomeric compositions (ee) of 2-bromoalkanes, which were enantioselectively enclathrated in (−)-tri-*o*-thymotide crystals, were determined via their conversion (with inversion of configuration) to 2-aminoalkanes and subsequent gas chromatographic analysis of TFA derivatives on the chiral diamide phase **4** (Arad-Yellin et al., 1980).

α-Amino acids and α-hydroxy acids are versatile starting materials for conversion to optically active building blocks (such as oxiranes) for chiral synthesis (Seuring and Seebach, 1977). Knowledge of the enantiomeric purity of compounds belonging to the "chiral carbon pool" is therefore essential. For instance, commercial (*S*)-ethyl lactate has been found to contain $1.67 \pm 0.06\%$ of the *R*-antipode as determined on **9** after derivatization with *tert*-butyl isocyanate (Hintzer et al., 1982).

One area of contemporary research in asymmetric catalysis is concerned with the homogeneous hydrogenation of dehydroamino acids to optically active amino acids with chiral rhodium(I)–phosphine coordination compounds (Kagan, 1975). In one instance, that of *N*-acetylphenylalanine, the enantiomeric bias achieved during asymmetric catalysis was underestimated due to an error in the reported specific rotation for this amino acid. Gas chromatographic analysis of *N*-acetylphenylalanine after derivatization with diazomethane on **4** not only established the correct enantiomeric yield obtained by asymmetric catalysis, but also permitted the extrapolation to the correct specific rotation $[\alpha_{max}]$ (Dang et al., 1975; Gelbard et al., 1976). Since relatively low enantiomeric yields were observed during the first asymmetric catalytic hydrogenations, the determination of ee on crude reaction mixtures by polarimetry appeared to be permissible (Knowles et al., 1972). Polarimetry should be replaced by gas chromatography, however, in order to confirm reports of enantiomeric yields >95% in more recent, highly sophisticated, enantioselective catalytic transformations (Knowles et al., 1975; Fryzuk and Bosnich, 1977; Brunner and Pieronczyk, 1979). Gas chromatography as a precise method of determining the ee of amino acids in asymmetric hydrogenation has been employed by Brown and Murrer (1980) and Scott et al. (1981).

Investigations of the selective synthesis of dipeptides by "double asymmetric induction" (Poulin et al., 1980) preferentially involves the determination of the diastereomeric and enantiomeric ratio of the four configurational isomers (i.e., *RR'/SS'* and *RS'/SR'*). The gas chromatographic separation of dipeptides as two pairs of enantiomers on a chiral stationary phase, as described by Ôi et al. (1980a), can be used in this area of asymmetric catalysis.

The availability of highly efficient chiral stationary phases for the quantitative resolution of acids, sugars, amines, alcohols, and related classes of compounds may, in the future, lead to an increased use of gas chromatography for the precise determination of high enantiomeric purities (99.9% > ee > 95%) of products obtained by asymmetric synthesis, kinetic resolutions, or enzymatic reactions.

C. Correlation of Absolute Configuration

The comparison of the sign of optical rotation for an optically active substrate with that of a reference compound of known chirality and identical structure permits the assignment of absolute configuration by polarimetry. The *direct* identification of absolute configuration by gas chromatography involves coinjection of solute and reference compound and inspection of the order of elution from a resolving chiral stationary phase. An *indirect* assignment of absolute configuration is based on the comparison of the order of peak emergence from a chiral stationary phase between the solute and a structurally related reference compound of known chirality.

The assignment of molecular configurations by gas chromatography is independent of chiroptical properties such as optical rotatory power and circular dichroism (CD). The sample size requirement is extremely low (10^{-8} g or less). The auxiliary resolving agent need not be enantiomerically pure, nor must its absolute configuration be known. In general, peak assignment can be carried out on partially resolved mixtures. It suffices for the reference compound to be enantiomerically enriched. The addition of a small amount of antipode (or of racemate) is essential for peak identification when the solute and reference compound are enantiomerically pure.

In the early fundamental investigation of Gil-Av *et al.* (1966) on the resolution of amino acids using optically active stationary phase (1), it was noted that D-amino acids eluted before the L-isomers when the resolving agent had the L-configuration. This order of elution has also been confirmed for the diamide phase of Feibush (1971) obtained from L-valine in either its low-molecular-weight (4) or polymeric forms (6–9) (Frank *et al.*, 1977; Saad *et al.*, 1979; Koenig *et al.*, 1981a). It has been found, however, that β- and γ-amino acids emerge in the reverse order, that is, D-enantiomers elute after the L-antipodes, from the diamide phase *N*-lauroyl-L-valine-6-undecylamide (Beitler and Feibush, 1976). A systematic study involving correlation of the order of chromatographic elution and molecular configuration revealed that *N*-TFA-α-, *N*-TFA-β-, and *N*-TFA-γ-amino acid esters elute from the chiral stationary phase carbonylbis(*N*-L-valine isopropyl ester) (3) in an order related to the steric requirements of the ligands surrounding the chiral carbon atom (Feibush *et al.*, 1972). If the solute molecule is viewed from the chiral carbon atom in the direction of the trifluoroacetylated nitrogen atom, the remaining three substituents will be arranged, according to decreasing relative size, in either a clockwise or an anticlockwise order. The enantiomers with a clockwise arrangement of the substituents have been found to elute as the second peak on **3** derived from L-valine. Thus, whereas for all N-TFA-alanine alkyl esters the D-enantiomer eluted before the L-antipode, the reverse was true for methyl-N-TFA leucinates because of the inverse steric requirements of the R and CO_2R' substituents. Furthermore, there was a lowering of resolvability as the effective sizes of CO_2R and R became similar, whereas resolution improved as the disparity between the relative sizes of the substituents increased.

A consistent relationship between configuration and elution order on **3** has also been observed for 2-aminoalkanes, making it possible to clarify prior confusion in the literature on configuration assignment caused by the effect of polar solvents on the sign of the optical rotation (Rubinstein *et al.*, 1973).

III. Direct Resolution of Underivatized Enantiomers on Metal-Containing Chiral Stationary Phases by Complexation Gas Chromatography

A. Methods

Experiments on the use of chiral metal coordination compounds as enantiospecific stationary phases for the resolution of enantiomers by complexation gas chromatography date back more than a decade. The first such chiral phase, a dicarbonylrhodium(I)-β-diketonate (**10**, Scheme 3) containing 3-trifluoroacetyl (1R)-camphorate was synthesized for the attempted resolution of chiral olefins (Schurig and Gil-Av, 1971; Schurig, 1972). The demonstration of chiral recognition by gas chromatographic solute–solvent interactions through coordination was delayed until 1977, when the first quantitative enantiomer resolution of racemic 3-methylcyclopentene (**15**) on the rhodium(I) chelate **10** (dissolved in squalane, $C_{30}H_{62}$, and coated onto a 200 m × 0.5 mm stainless steel capillary column) was reported (Schurig, 1977; Schurig and Gil-Av, 1976–1977) (Fig. 4). Thus, although further efforts to widen the scope of enantiomer resolution to other racemic olefins failed, it was established for the first time that complexation gas chromatography may exhibit the necessary thermodynamic and kinetic parameters for enantiomer discrimination via coordination (Schurig, 1980).

Following the first reports, Golding *et al.* (1977) used the chelate **11**, which had been shown to be a powerful chiral NMR shift reagent (Whitesides and Lewis, 1970), for the semiquantitative gas chromatographic resolution of methyloxirane. Schurig and Bürkle (1978) quantitatively resolved methyloxirane and *trans*-2,3-dimethyloxirane on nickel(II)–bis[3-trifluoroacetyl (1R)-camphorate]

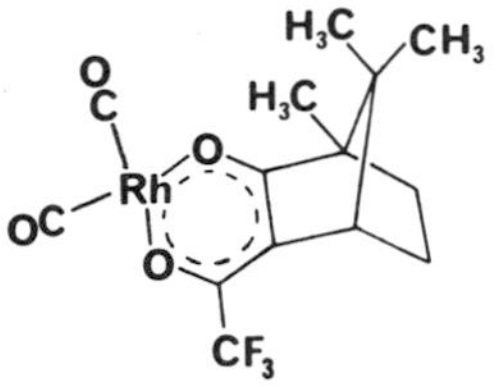

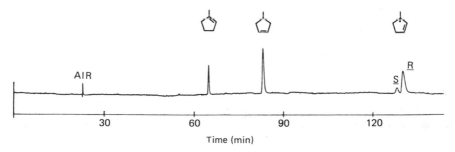

10

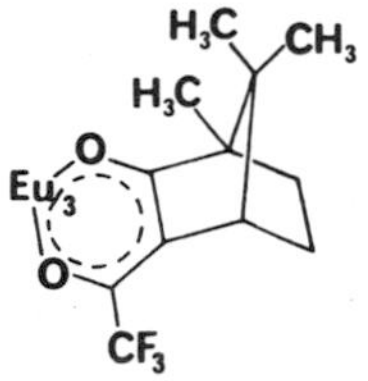

11

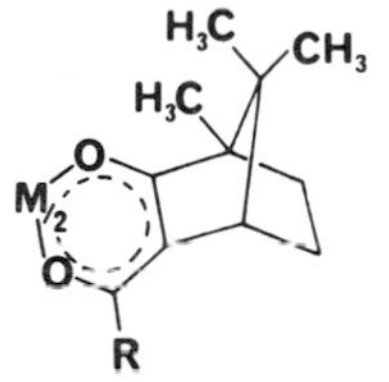

12a M = Ni; R = CF$_3$
12b M = Ni; R = C$_3$F$_7$
13 M = Mn; R = C$_3$F$_7$

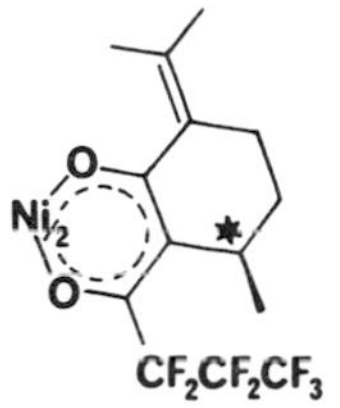

14

Scheme 3. Chiral metal chelates for enantiomer resolution by complexation gas chromatography.

(**12a**) dissolved in squalane and coated onto a 200 m × 0.5 mm nickel capillary column.

These results have prompted further investigations of the resolution of oxygen-, nitrogen-, and sulfur-containing δ-donor solutes such as cyclic ethers (up to six-membered rings), 1-chloroaziridines, thiiranes, thietanes, cyclic ketones, and aliphatic alcohols on 3-heptafluorobutanoyl (1R)-camphorates of nickel(II) (**11b**) as shown in Fig. 5. Table II lists separation factors α for the resolution of racemic solutes that have been quantitatively resolved by complexation gas chromatography.

Fig. 4. Gas chromatographic separation of isomeric methylcyclopentenes including enantiomer resolution of 3-methylcyclopentene (enriched in the R-antipode) on a stainless steel capillary coated with **10** in squalane at 20°C. From Schurig and Gil-Av (1976–77).

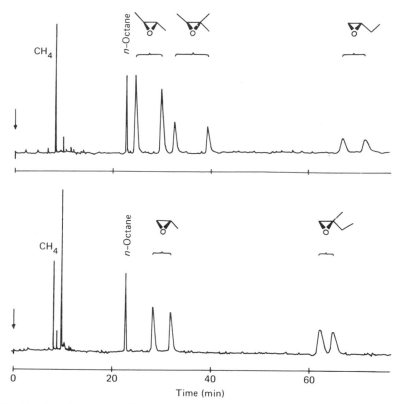

Fig. 5. Gas chromatographic separation of enantiomers of mono-, di-, and trialkyl-substituted oxiranes on a nickel capillary (100 m × 0.5 mm) coated with **12b** (0.1 *m* in squalane) at 60°C. V. Schurig and W. Bürkle (unpublished).

It should be mentioned that enantiomer resolution by complexation gas chromatography is not restricted to cyclic compounds, as the quantitative resolution of *tert*-butylmethylcarbinol or the pheromone sulcatol demonstrates (Weber, 1979; V. Schurig *et al.*, unpublished). Noteworthy is the resolvability on a nickel(II) chelate derived from 5-heptafluorobutanoyl-(*R*)-pulegone (**14**) of various spiroketals (**16**, Scheme 4) (Weber *et al.*, 1980), the bicyclic pheromone ketals *exo*- and *endo*-brevicomin and frontalin (**17**) (Schurig *et al.*, 1983), and the tricyclic acetal lineatin (**18**) (Weber and Schurig, 1981). The direct resolution of the invertomers of 1-chloro-2-dimethylaziridine (**19**; α = 1.5°, 60°C) on **12b** deserves special mention because it demonstrates the configurational stability of the three-coordinated nonplanar nitrogen atom (Schurig *et al.*, 1979).

A further improvement of resolution efficiency and speed of analysis has been effected by the use of pretreated glass capillary columns (Duran) coated with **13**

TABLE II

Separation Factor α of Enantiomer Resolution for Racemic Underivatized Solutes
by Complexation Gas Chromatography on Optically Active Metal Chelates

Solute	Metal chelate	Temp. (°C)	α^a	Reference
Methyloxirane	**12b**	60	1.19	Schurig and Bürkle (1982)
Ethyloxirane	**13**	60	1.15	Schurig and Weber (1981)
Isopropyloxirane	**13**	60	1.17	Schurig and Weber (1981)
tert-Butyloxirane	**13**	60	1.15	Schurig and Weber (1981)
sec-Butyloxirane (erythro)	**13**	60	1.07	Schurig and Weber (1981)
sec-Butyloxirane (threo)	**13**	60	1.19	Schurig and Weber (1981)
2-Ethyl-2-methyloxirane	**12b**	60	1.05	Schurig and Bürkle (1982)
trans-2,3-Dimethyloxirane	**12b**	60	1.32	Schurig and Bürkle (1982)
cis-2-Ethyl-3-methyloxirane	**13**	60	1.04	Schurig and Weber (1981)
Trimethyloxirane	**12b**	60	1.30	Schurig and Bürkle (1982)
2-Methyloxetane	**13**	60	1.11	Schurig and Weber (1981)
trans-2,3-Dimethyloxetane	**12b**	70	1.21	Schurig and Bürkle (1982)
2-Methyltetrahydrofuran	b	60	1.09	Schurig and Weber (1981)
3-Methyltetrahydrofuran	b	60	1.06	Schurig and Weber (1981)
trans-2,5-Dimethyltetrahydrofuran	**12b**	60	1.18	Schurig and Bürkle (1982)
2-Methyltetrahydropyran	**12b**	60	1.05	Schurig and Bürkle (1982)
Epichlorhydrine	**12b**	70	1.16	Schurig and Bürkle (1982)
Methylthiirane	**12b**	60	1.05	Schurig and Bürkle (1982)
Ethylthiirane	**12b**	60	1.08	Schurig and Bürkle (1982)
trans-2,3-Dimethylthiirane	**12b**	70	1.07	Schurig and Bürkle (1982)
trans-2,4-Dimethylthietane	**12b**	70	1.05	Schurig and Bürkle (1982)
1-Chloro-2,2-dimethylaziridine	**12b**	60	1.50	Schurig et al. (1979)
trans-1-Chloro-2-methylaziridine	**12b**	63	1.29	Schurig and Bürkle (1982)
1,6-Dioxaspiro[4.4]nonane	**12b**	70	1.05	Schurig and Bürkle (1982)
trans-2,5-Dimethylcyclopentanone	**12b**	70	1.10	Schurig and Bürkle (1982)
tert-Butylmethylcarbinol	**12b**	70	1.09	Schurig and Bürkle (1982)
4-Methyl-3-heptanol (erythro)	**13**	65	1.04	V. Schurig et al. (unpubl.)
4-Methyl-3-heptanol (threo)	**12b**	90	1.03	V. Schurig et al. (unpubl.)
6-Methyl-5-hepten-2-ol (sulcatol)	**14**	60	1.07	V. Schurig et al. (unpubl.)

[a]Retention time t' (adjusted for the dead volume) of the second eluting enantiometer over that of the first eluting enantiomer.

[b]Manganese (II)–bis[3-pentafluorobenzoyl (1R)-camphorate].

dissolved in polysiloxane OV-101 (Fig. 6) (Schurig et al., 1983). Short packed columns have also been employed for the resolution of alkyloxiranes and **18** by complexation gas chromatography (Schurig and Weber, 1981). Ôi et al. (1980b, 1981b) resolved α-hydroxy acid esters and amino alcohols on a chiral copper(III)–Schiff base chelate. Although separation factors α were high for certain compounds, the peak resolution was somewhat diminished due to poor column efficiency.

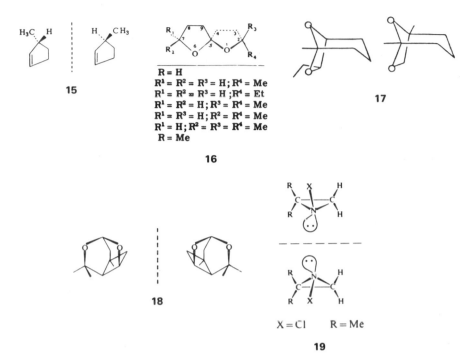

Scheme 4. Examples of chiral molecules quantitatively resolvable by complexation gas chromatography on optically active metal chelates.

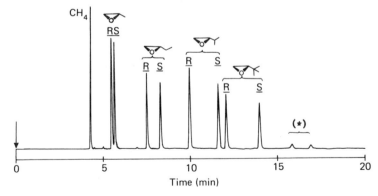

Fig. 6. Gas chromotographic separation of the enantiomers of alkyloxiranes on a glass capillary coated with **13** in OV-101 (42 m × 0.25 mm) at 40°C. V. Schurig and R. Weber (unpublished). Star denotes chiral impurity of *tert*-butyloxirane.

It is important to note that, in general, enantiomer resolution by complexation gas chromatography does not require solute derivatization. Because the exploitation of this technique has only begun it is likely that racemic solutes bearing chemical functionalities not tested thus far will be resolved in the future. An improvement of separation factors α may be achieved by the use of "tailored" metal chelates as enantiospecific stationary phases.

B. Applications

Complexation gas chromatography can be employed as a useful technique for the determination of enantiomeric compositions in asymmetric synthesis. The merits of the method are as follows:

1. No substrate derivatization and no work-up procedures (isolation and purification, which may lead to accidental enantiomer enrichment) are required.
2. The size of sample required is very low (10^{-8} g) and may even be decreased by the use of selective detection devices.
3. The sample can be directly collected from the vapor phase of the reaction mixture (this significant simplification of analysis requires the absence of diastereoselective effects in enantiomerically enriched mixtures).

The technique of complexation gas chromatography is especially suitable for trace analysis in asymmetric synthesis. Because peak resolution is generally rapid and no physical or chemical manipulation of the substrate is required, the enantiomeric yield of a chiral transformation can be monitored during the entire course of the reaction. The knowledge of the enantiomeric composition (ee) of the product formed at the very beginning of an asymmetric reaction is particularly important for rationalizing mechanisms of chiral recognition. Moreover, the continuous monitoring of ee during the entire reaction period is desirable if the chiral reaction product may undergo racemization [as, e.g., in the case of 2-phenylpropanal formed by rhodium-catalyzed asymmetric hydroformylation of styrene (Botteghi *et al.,* 1974)] or if the chiral auxiliary undergoes structural changes during the course of the reaction. The continuous monitoring of enantiomeric compositions is required per se when correlations between ee, the difference of relative reaction rates, and the extent of conversion in kinetic resolutions are to be established quantitatively (Balavoine *et al.,* 1974; Martin *et al.,* 1981; Sepulchré *et al.,* 1981).

The first quantitative enantiomer resolution of an *underivatized* substrate was achieved for 3-methylcyclopentene (**15**) on the rhodium chelate **10** by complexation gas chromatography (Schurig, 1977). Relevant applications of the gas chromatographic technique for **15** involved the detection of 14% inversion of configuration (remaining constant during the entire reaction period), coformation of 4-methylcyclopentene, and rearrangement to 1-methylcyclopentene in the preparation of (*R*)-**15** by acid dehydration of (*R*)-3-methylcyclopentanol (*cis* and *trans*) (see Fig. 4) as well as evidence for the absence of racemization and rearrangement by acetate pyrolysis at 530°C of epimeric alcohols to (*R*)-**15** (Schurig and Gil-Av, 1976–1977). The precise determination of ee and the amounts of alkene isomers permitted the interpretation of the CD spectrum (Levi *et al.*, 1980) and the extrapolation of the specific rotation of (*R*)-**15**, $[\alpha]_D^{20}$ + 174.5 ± 4° (Schurig and Gil-Av, 1976–1977), this value being more than double that previously reported (Mousseron *et al.*, 1946) and calculated (Brewster, 1959). The new figure has been found to be consistent with a correlation between the $[\alpha]_D$ of 3-substituted cyclopentenes and their inverse molecular weight (Chapman *et al.*, 1978).

In Fig. 7 preliminary results on the kinetic resolution of racemic 3-methylcyclopentene (**15**) by hydrogenation with Rh(PPh₃)₃Cl/(−)-DIOP (Kagan, 1975) are shown (V. Schurig, unpublished). Thus, with the DIOP catalyst derived from (+)-tartaric acid a preferential consumption of the *R*-antipode is readily observed.

The quantitative resolvability of underivatized alkyloxiranes by complexation gas chromatography (Schurig *et al.*, 1978) has greatly facilitated the direct determination of enantiomeric yields in the small-scale asymmetric epoxidation of unfunctionalized simple olefins with the reagent molybdenum(VI)–oxodiperoxo-(*S*)-dimethyllactamide (**20**) (Kagan *et al.*, 1979). The methodology also permitted the continuous kinetic monitoring of the asymmetric epoxidation reaction. Thus, the ee of the oxiranes remained constant during the entire course of the reaction (Mark, 1980) (Fig. 8). This result proved that the formation of the oxiranes was indeed asymmetrically induced and that enantiomer enrichment

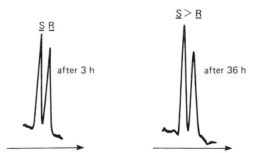

Fig. 7. Kinetic resolution of 3-methylcyclopentene (**15**) by hydrogenation with Rh(PPh₃)₃Cl/(−)-DIOP in benzene at 20°C and 1 bar. For gas chromatographic conditions see legend of Fig. 4. V. Schurig (unpublished result).

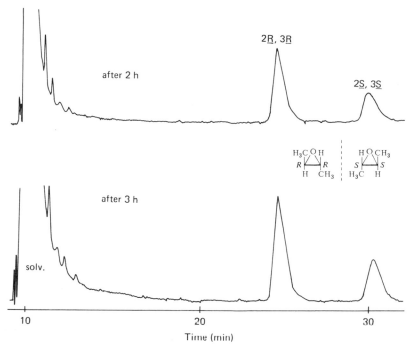

Fig. 8. Enantioselective epoxidation of *trans*-2-butene with Mo(O)(O₂)₂-(S)-di-methyllactamide (**20**) to (2R,3R)-*trans*-dimethyloxirane (ee = 35 ± 1%) in nitrobenzene at 20°C and 1 bar. For gas chromatographic conditions see legend of Fig. 5. From Mark (1980).

was not due to kinetic resolution of racemic oxirane while undergoing slow decomposition. The resolvability of the four configurational isomers of *sec*-butyloxirane (Koppenhoefer *et al.*, 1982) (Fig. 9, left) permitted the study of "double asymmetric induction" in the epoxidation of 3-methyl-1-pentene with **20**. Although asymmetric induction was low at 22°C, (2R,3R)-*sec*-butyloxirane was formed preferentially relative to other isomers (Fig. 9, right) (V. Schurig and H. B. Kagan, unpublished).

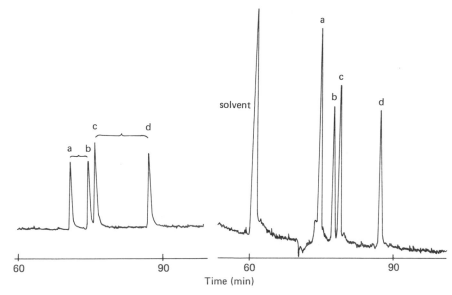

Fig. 9. Enantioselective epoxidation of racemic 3-methyl-1-pentene with Mo(O)-(O₂)₂-(S)-dimethyllactamide (**20**) to *erythro* and *threo-sec*-butyl-oxirane in nitrobenzene at 20°C and 1 bar. Peak assignment: (a) 2R,3R; (b) 2S,3S; (c) 2R,3S; (d) 2S,3R. Left: racemic oxiranes; right: reaction mixture. From V. Schurig and H. B. Kagan (unpublished result).

It has also been shown that (2R,3R)-*trans*-dimethyloxirane is produced with 65% ee at the initial stages of the epoxidation of *trans*-2-butene with the chiral reagent molybdenum(VI)–oxodiperoxo-(2S,3S)-butanediol (Mark, 1980). Because the ee of the oxirane rapidly decreased, the asymmetric bias observed at the beginning of the synthesis would have escaped detection had measurements been performed only after completion of the reaction, as is usually practiced.

The ee of trace amounts of methyloxirane, 2,3-dimethyloxirane, 2,3-dimethylthiirane, and 2-methyltetrahydrofuran, enantioselectively enclathrated in (+)-tri-*o*-thymotide crystals, has been determined by complexation gas chromatography on the nickel chelate **12b**.

Finally, by combining the two gas chromatographic methods of enantiomer resolution described in Sections II,A and III,A, precise enantiomeric purities were determined for α-hydroxy and α-amino acids ("chiral carbon pool" molecules) and also for products of their chemical transformations, such as oxiranes. It was shown that various transformations, including substitutions proceeding with retention and inversion of configuration at the chiral center, were essentially free of racemization (Schurig *et al.*, 1978; Hintzer *et al.*, 1981; Koppenhoefer *et al.*, 1982).

The precise determination of chemical yields in small-scale asymmetric reactions is carried out by adding a predetermined amount of the enantiomerically pure antipode as an internal standard (exhibiting identical physical and chemical properties) to the reaction mixture and recording the change of the enantiomeric composition of the chiral product. Quantification by this technique, referred to as *enantiomer labeling* (Bonner, 1973; Frank *et al.*, 1978c), is free of errors due to incomplete sample collection, losses during injection, and differing detector response.

C. Correlation of Absolute Configuration

Schurig *et al.* (1978) extended the application of gas chromatography to the determination of absolute configurations (Section II,C) to include enantiospecific solute–solvent interactions through coordination. Thus an empirical quadrant rule, predicting the enantiomer elution order of chiral three-membered-ring heterocycles of structure **21** from nickel(II)–bis[3-heptafluorobutanoyl (1*R*)-cam-

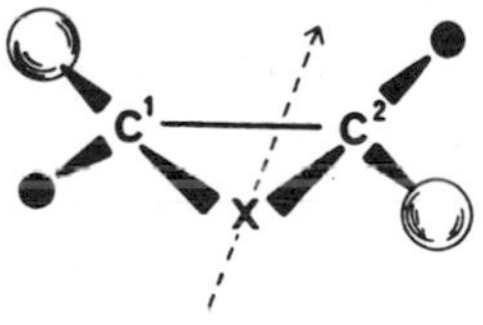

21　X = NH, O, S

phorate], (1*R*)-**12b**, has been formulated. When the solute molecule is viewed from the donor heteroatom in the direction of the horizontal C—C bond, the absolute configuration of the enantiomer eluting as the second peak (corresponding to a stronger interaction) is that in which the bulkier group is situated on the upper left at C-1 and/or lower right at C-2. Thus, it has now been established (Schurig and Bürkle, 1982) that (*S*)-methyl, (*S*)-ethyl-, (*S*)-isopropyl-, (*S*)-*sec*-butyl- (two isomers), and (*S*)-*tert*-butyloxirane as well as (2*S*,3*S*)-*trans*-dimethyloxirane, (*S*) trimethyloxirane, (1*S*,2*S*)-1-chloro-2-methylaziridine, and (*S*)-methylthiirane elute as the second peak on **12b** or **13** derived from (+)-(1*R*)-camphor. A notable exception to the quadrant rule has been detected, however, for (2*S*,3*S*)-*trans*-dimethyloxirane of identical stereochemistry (Schurig and Bürkle, 1982). Thus, caution must be exercised in predicting configuration, even in systems of apparent simplicity.

The absolute configuration of the predominant oxirane enantiomers obtained by asymmetric epoxidation of olefins with **20** have been deduced by the quadrant rule and are supported by chemical evidence (Kagan *et al.*, 1979).

IV. Precision of Analysis and Practical Considerations

Bonner *et al.* (1974) compared the accuracy and precision of gas chromatography for the determination of enantiomeric compositions of leucine. The method showed comparable accuracy (i.e., 0.03–0.7% absolute error) and reproducibility (i.e., 0.03–0.6% standard deviation) in the range 0–80% ee. Digital electronic integration of a 50/50 racemic mixture gave standard deviations of 50 ± 0.03% (four measurements) and 50 ± 0.17% (three measurements).

As mentioned before, gas chromatography is the method of choice for the precise determination of ee > 95%. Thus, commercial (S)-ethyl lactate (Fluka, labeled $[\alpha]_D^{20}$ −11.0 ± 0.3°) contained 1.673 ± 0.065% of the R-antipode (ee = 96.65 ± 0.13%), as revealed by seven measurements using the *tert*-butylurethane on Chirasil-Val (**6**) (Hintzer *et al.*, 1982).

In Fig. 10 the complexation gas chromatogram of (R)-isopropyloxirane obtained from (S)-valine (ee > 99.5%) (Koppenhoefer *et al.*, 1982) and resolved on **13** is shown. Digital electronic integration revealed 1.04 ± 0.02% (ee = 97.92 ± 0.04%; five measurements for baseline-separated peaks) and 1.12 ± 0.04% S-enantiomer (ee = 97.76 ± 0.08%; nine measurements for nearly baseline-separated peaks evaluated by the trapezoidal mode of integration). The determination of ee by the unconventional method of Xeroxing the (expanded) chromatogram, cutting the peak areas, and comparing the weight of the sum and

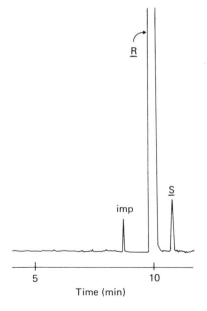

Fig. 10. Gas chromatographic determination of the enantiomeric purity (ee) of (R)-isopropyloxirane prepared from (S)-valine. (Koppenhoefer *et al.*, 1982). For gas chromatographic conditions see legend of Fig. 6 (imp denotes impurity) (V. Schurig and R. Weber, unpublished).

the difference of the peak areas gave ee 97.2 ± 0.5%. It is obvious from an inspection of Fig. 10 that even traces of enantiomeric impurities can be detected and accurately determined by gas chromatography.

The nondigital evaluation of the peak areas of the chromatographic eluates shown in Fig. 4 permitted the extrapolation of the absolute specific rotation for 3-methylcyclopentene (**15**) (i.e., $[\alpha_{max}]_D^{20}$ +174.5° with a standard deviation of ±4°) (Schurig and Gil-Av, 1976–77).

Reversal of the elution pattern for enantiomerically enriched solutes by inversion of the chirality of the optically active stationary phase (Schurig, 1977; Schurig and Gil-Av, 1976–77) can be employed to differentiate a true enantiomer resolution from accidental peak splittings due to the presence of accompanying impurities. This technique has also been suggested for the determination of minute amounts of enantiomeric impurities if the deficient enantiomer elutes *after* the main component and may thus be lost in the tail of the abundant enantiomer (Bonner and Blair, 1979).

Gas chromatographic high-resolution capillary columns coated with **1–14** exhibit extended lifetimes. The pretreatment of glass capillaries (Nicholson *et al.*, 1979) and the deactivation of metal capillaries (Schurig, 1980) require some experience. When separation factors are greater than 1.2, the use of packed columns may be useful. To improve enantiomer resolution the lowest possible temperature should be employed. Traces of water and oxygen should be absent from the carrier gas (helium, argon, or nitrogen), and the amount of sample injected should be minimized. The use of capillary columns requires sample splitting in the injector and the absence of dead volumes in the flow system.

V. Conclusions

Gas chromatography is a highly efficient and precise technique for determining the enantiomeric composition (ee) of chiral solutes. The method is especially suitable for the determination of small (i.e., ee ≈ 0%) and high (i.e., ee > 95%) enantiomeric purities. The accuracy of ee when measured by digital peak integration can be as small as ±0.1%. Restrictions of the method are due to the requirements of solute volatility, thermal stability, and quantitative resolvability.

Acknowledgments

Support for this work from Deutsche Forschungsgemeinschaft and Fonds der chemischen Industrie is gratefully acknowledged. The author thanks his co-workers W. Bürkle, K. Hintzer, and R. Weber for their invaluable contributions to complexation gas chromatography.

References

Arad-Yellin, R., Green, B. S., and Knossow, M. (1980). *J. Am. Chem. Soc.* **102**, 1157.
Balavoine, G., Moradpour, A., and Kagan, H. B. (1974). *J. Am. Chem. Soc.* **96**, 5152.
Bayer, E., and Frank, H. (1980). *ACS Symp. Ser.* No. 121.
Bayer, E., Gil-Av, E., Koenig, W. A., Nakaparksin, S., Oró, J., and Parr, W. (1970). *J. Am. Chem. Soc.* **92**, 1738.
Beitler, U., and Feibush, B. (1976). *J. Chromatogr.* **123**, 149.
Bonner, W. A. (1973). *J. Chromatogr. Sci.* **11**, 101.
Bonner, W. A. (1979). *In* "Origins of Optical Activity in Nature" (D. C. Walker, ed.), p. 5. Elsevier, Amsterdam.
Bonner, W. A., and Blair, N. E. (1979). *J. Chromatogr.* **169**, 153.
Bonner, W. A., Van Dort, M. A., and Flores, J. J. (1974). *Anal. Chem.* **46**, 2104.
Botteghi, C., Consiglio, G., and Pino, P. (1974). *Liebigs. Ann. Chem.*, p. 864.
Brewster, J. H. (1959). *J. Am. Chem. Soc.* **81**, 5493.
Brown, J. M., and Murrer, B. A. (1980). *Tetrahedron Lett.*, p. 581.
Brunner, H., and Pieronczyk, W. (1979). *Angew. Chem. Int. Ed. Engl.* **18**, 620.
Chang, S.-C., Charles, R., and Gil-Av, E. (1980). *J. Chromatogr.* **202**, 247.
Chang, S.-C., Charles, R., and Gil-Av, E. (1982). *J. Chromatogr.* **238**, 29.
Chapman, O. L., Mattes, K. C., Sheridan, R. S., and Klun, J. A. (1978). *J. Am. Chem. Soc.* **100**, 4878.
Charles, R., and Gil-Av, E. (1980). *J. Chromatogr.* **195**, 317.
Charles, R., Beitler, U., Feibush, B., and Gil-Av, E. (1975). *J. Chromatogr.* **112**, 121.
Coleman, C. B., Cooper, G. D., and O'Donnell, J. F. (1966). *J. Org. Chem.* **31**, 975.
Dang, T. P., Poulin, J.-C., and Kagan, H. B. (1975). *J. Organomet. Chem.* **91**, 105.
Feibush, B. (1971). *J. Chem. Soc. Chem. Comm.*, p. 544.
Feibush, B., and Gil-Av, E. (1967). *J. Gas Chromatogr.* **5**, 257.
Feibush, B., and Gil-Av, E. (1970). *Tetrahedron* **26**, 1361.
Feibush, E., Gil-Av, E., and Tamari, T. (1970). *Isr. J. Chem.* **8**, 50.
Feibush, E., Gil-Av, E., and Tamari, T. (1972). *J. Chem. Soc. Perkin Trans.* 2, p. 1197.
Frank, H., Nicholson, G. J., and Bayer, E. (1977). *J. Chromatogr. Sci.* **15**, 174.
Frank, H., Nicholson, G. J., and Bayer, E. (1978a). *J. Chromatogr.* **146**, 197.
Frank, H., Nicholson, G. J., and Bayer, E. (1978b). *Angew. Chem. Int. Ed. Engl.* **17**, 363.
Frank, H., Nicholson, G. J., and Bayer, E. (1978c). *J. Chromatogr.* **167**, 187.
Frank, H., Woiwode, W., Nicholson, G. J., and Bayer, E. (1981). *Liebigs Ann. Chem.*, p. 354.
Fryzuk, M. D., and Bosnich, B. (1977). *J. Am. Chem. Soc.* **99**, 6262.
Gelbard, G., Kagan, H. B., and Stern, R. (1976). *Tetrahedron* **32**, 233.
Gil-Av, E. (1975). *J. Mol. Evol.* **6**, 131.
Gil-Av, E., and Feibush, B. (1967). *Tetrahedron Lett.*, p. 3345.
Gil-Av, E., and Nurok, D. (1974). *Adv. Chromatogr.* **10**, 99.
Gil-Av, E., Feibush, B., and Charles-Sigler, R. (1966). *Tetrahedron Lett.*, p. 1009.
Gil-Av, E., Feibush, B., and Charles-Sigler, R. (1967). *In* "Gas Chromatography 1967" (A. B. Littlewood, ed.), p. 227. Inst. Petroleum, London.
Golding, B. T., Sellars, P. J., and Wong, A. K. (1977). *J. Chem. Soc. Chem. Comm.*, p. 570.
Guetté, J. P., and Horeau, A. (1965). *Tetrahedron Lett.*, p. 3049.
Halpern, B. (1977). *In* "Handbook of Derivatives for Chromatography" (K. Blau and G. King, eds.), p. 457. Heyden, London.
Hintzer, K., Weber, R., and Schurig, V. (1981). *Tetrahedron Lett.*, p. 55.
Hintzer, K., Koppenhoefer, B., and Schurig, V. (1982). *J. Org. Chem.* **47**, 3850.

Horeau, A. (1969). *Tetrahedron Lett.*, p. 3121.

Kagan, H. B. (1975). *Pure Appl. Chem.* **43**, 401.

Kagan, H. B., Mimoun, H., Mark, C., and Schurig, V. (1979). *Angew. Chem. Int. Ed. Engl.* **18**, 485.

Karagounis, G., and Lemperle, E. (1962). *Fresenius Z. Anal. Chem.* **189**, 131.

Karagounis, G., and Lippold, G. (1959). *Naturwissenschaften* **46**, 145.

Knowles, W. S., Sabacky, M. J., and Vineyard, B. D. (1972). *J. Chem. Soc. Chem. Commun.*, p. 10.

Knowles, W. S., Sabacky, M. J., Vineyard, B. D., and Weinkauff, D. J. (1975). *J. Am. Chem. Soc.* **97**, 2567.

Koenig, W. A., and Benecke, I. (1981). *J. Chromatogr.* **209**, 91.

Koenig, W. A., Sievers, S., and Benecke, I. (1981a). *Proc. Int. Symp. Capillary Chromatogr. 4th,* p. 703.

Koenig, W. A., Benecke, I., Sievers, S. (1981b). *J. Chromatogr.* **217**, 71.

Koenig, W. A., Benecke, I., and Bretting, H. (1981c). *Angew. Chem. Int. Ed. Engl.* **20**, 693.

Koenig, W. A., Benecke, I., and Sievers, S. (1982a). *J. Chromatogr.* **238**, 427.

Koenig, W. A., Francke, W., and Benecke, I. (1982b). *J. Chromatogr.*, **239**, 227.

Koppenhoefer, B. (1980). Ph.D. Dissertation, Univ. of Tübingen.

Koppenhoefer, B., Weber, R., and Schurig, V. (1982). *Synthesis*, p. 316.

Levi, M., Cohen, D., Schurig, V., Basch, H., and Gedanken, H. (1980). *J. Am. Chem. Soc.* **102**, 6972.

Lochmüller, C. H., and Souter, R. W. (1975). *J. Chromatogr.* **113**, 283.

Mark, C. (1980). Ph.D. Dissertation, Univ. of Tübingen.

Martin, V. S., Woodard, S. S., Katsuki, T., Yamada, Y., Ikeda, M. and Sharpless, K. B. (1981). *J. Am. Chem. Soc.* **103**, 6237.

Mousseron, M., Richaud, R., and Granger, G. (1946). *Bull. Soc. Chim. Fr.*, p. 222.

Nakaparksin, S., Birell, P., Gil-Av, E., and Oró, J. (1970). *J. Chromatogr. Sci.* **8**, 177.

Nicholson, G. J., Frank, H., and Bayer, E. (1979). HRC CC *J. High Resolut. Chromatogr. Chromatogr. Commun.* **2**, 411.

Ôi, N., Horiba, M., Kitahara, H., and Shimada, H. (180a). *J. Chromatogr.* **202**, 302.

Ôi, N., Horiba, M., Kitahara, H., Doi, T., Tani, T., and Sakakibara, T. (1980b). *J. Chromatogr.* **202**, 305.

Ôi, N., Doi, T., Kitahara, H., and Inda, Y. (1981a). *J. Chromatogr.* **208**, 404.

Ôi, N., Shiba, K., Tani, T., Kitahara, H., and Doi, T. (1981b). *J. Chromatogr.* **211**, 274.

Oró, J., and Updegrove, W. S., Gilbert, J., McReynolds, J., Gil-Av, E., Ibanez, J., Zlatkis, A., Flory, D. A., Levy, R. L., and Wolf, C. (1970). *Science (Washington, D.C.)* **167**, 765.

Pandey, R. C., Meng, H., Cook, J. C., Jr., and Rinehart, K. L., Jr. (1977). *J. Am. Chem. Soc.* **99**, 5203, 5206, 8469.

Parr, W., Yang, C., Bayer, E., and Gil-Av, E. (1970). *J. Chromatogr. Sci.* **8**, 591.

Poulin, J. C., Meyer, D., and Kagan, H. B. (1980). *C. R. Séances Acad. Sci., Ser. C.* **291**, 69.

Raban, M., and Mislow, K. (1967). *Top. Stereochem.* **2**, 199.

Rubinstein, H., Feibush, B., and Gil-Av, E. (1973). *J. Chem. Soc. Perkin Trans. 2*, p. 2094.

Saeed, T., Sandra, P., and Verzele, M. (1979). *J. Chromatogr.* **186**, 611.

Saeed, T., Sandra, P., and Verzele, M. (1980). HRC CC *J. High Resolut. Chromatogr. Chromatogr. Commun.* **3**, 35.

Schurig, V. (1972). *Inorg. Chem.* **11**, 736.

Schurig, V. (1977). *Angew. Chem. Int. Ed. Engl.* **16**, 110.

Schurig, V. (1980). *Chromatographia* **13**, 263.

Schurig, V., and Bürkle, W. (1978). *Angew. Chem. Int. Ed. Engl.* **17**, 132.

Schurig, V., and Bürkle, W. (1982). *J. Am. Chem. Soc.* **104**, 7573.

Schurig, V., and Gil-Av, E. (1971). *J. Chem. Soc. Chem. Commun.*, p. 650.

Schurig, V., and Gil-Av, E. (1976–77). *Isr. J. Chem.* **15,** 96.

Schurig, V., and Weber, R. (1981). *J. Chromatogr.* **217,** 51.

Schurig, V., and Weber, R. (1983). *Angew. Chem. Int. Ed. Engl.*, accepted for publication.

Schurig, V., Koppenhoefer, B., and Bürkle, W. (1978). *Angew. Chem. Int. Ed. Engl.* **17,** 937.

Schurig, V., Bürkle, W., Zlatkis, A., and Poole, C. F. (1979). *Naturwissenschaften* **66,** 423.

Schurig, V., Koppenhoefer, B., and Bürkle, W. (1980). *J. Org. Chem.* **45,** 538.

Schurig, V., Weber, R., Nicholson, G. J., Oehlschlager, A. C., Pierce, H., Jr., Pierce, A. M.,
Borden, J. H., and Rylxer, L. C. (1983). *Naturwissenschaften* **70,** 92.

Scott, J. W., Keith, D. D., Nix, G., Jr., Parrish, D. R., Remington, S., Roth, G. P., Townsend, J.
M., Valentine, D., Jr., and Yang, R. (1981). *J. Org. Chem.* **46,** 5086.

Sepulchré, M., Spassky, N., Mark, C., and Schurig, V. (1981). *Makromol. Chem. Rapid Commun.*
2, 261.

Seuring, B., and Seebach, D. (1977). *Helv. Chim. Acta* **60,** 1175.

Sullivan, G. R. (1978). *Top. Stereochem.* **10,** 287.

Weber, R. (1979). Diplom. Thesis, Univ. of Tübingen.

Weber, R., and Schurig, V. (1981). *Naturwissenschaften* **68,** 330.

Weber, R. Hintzer, K., and Schurig, V. (1980). *Naturwissenschaften* **67,** 453.

Weinstein, S., Feibush, B., and Gil-Av, E. (1976). *J. Chromatogr.* **126,** 97.

Whitesides, G. M., and Lewis, D. W. (1970). *J. Am. Chem. Soc.* **92,** 6979.

6

Separation of Enantiomers by Liquid Chromatographic Methods

William H. Pirkle
John Finn
Roger Adams Laboratory
School of Chemical Sciences
University of Illinois
Urbana, Illinois

I. Introduction and Background

Within the next few years, there will be a dramatic increase in the solution of stereochemical problems by liquid chromatographic (LC) means. The great extent to which the use of advanced-level LC apparatus is becoming routine augurs well for the general acceptance of rapid, inexpensive, and operationally simple LC methods for determining enantiomeric purity, determining absolute configuration, or actually separating useful quantities of stereoisomers. The development of the stereochemical methodology necessary for the solution of these problems is actively underway, recent findings clearly indicating the power and widespread potential of the approach. Even at this early stage of development, it is possible to assess enantiomeric purity accurately for thousands of compounds using analytical high-pressure liquid chromatographic (HPLC) technology. In many such instances absolute configurations are determined concurrently. Whereas analytical-scale technology make possible the separation of stereoisomers in milligram quantities, gram-scale separations can be effected on preparative medium-pressure liquid chromatographic (MPLC) systems, which could easily be automated. A major advantage of chromatographic separations relative to the more classical separation method of fractional crystallization is that, for a ''first-time'' separation, the outcome of the former is frequently predictable, whereas that of the latter is not.

As mentioned, increasing effort is being devoted to the development of LC methods for the separation of stereoisomers; several reviews, by Buss and Vermeulen (1968), Lochmüller and Souter (1975), Krull (1977), Tamegai et al. (1979), Audebert (1979), Blaschke (1980), and Davanakov (1980), have provided comprehensive coverage of these methods. This chapter stresses methods of proven utility and endeavors to delineate the factors responsible for the success of the method. An understanding of the principles underlying these separations should allow a potential user to select (or reject) rationally a method as being applicable (or inapplicable) to a particular problem.

The separation of enantiomers requires the intervention of a chiral agent. This can take the form of long-term derivatization of a pair of enantiomers with a chiral derivatizing agent (CDA) to afford chromatographically separable diastereomers, or it can take the form of short-term interaction of the enantiomers with a chiral agent to afford short-lived diastereomeric complexes. Because the first approach requires a discrete derivatization step, we term this approach *indirect* as opposed to the *direct* resolutions afforded by the second approach. If, during the direct approach, the chiral agent is the column itself [i.e., one is using a column packed with a chiral stationary phase (CSP)], then the intent is that the diastereomeric complexes will be of nonidentical stability and that the enantiomers will elute at different times. Alternatively, one might add a chiral agent to the mobile phase with the intention that the diastereomeric complexes will differ either in stability or chromatographic behavior on an achiral column. Each of these three techniques has its own advantages and disadvantages.

At the outset we would like to define a few chromatographic terms and concepts for readers who are not yet experienced chromatographers. The volume of solvent required to elute a *nonretained* solute is equal to the void volume of the column. An additional volume of solvent will then be required to elute a *retained* solute. The ratio of this additional volume of solvent to the void volume is termed the capacity ratio κ. The separability factor α for two solutes is κ_2/κ_1, a value that corresponds to the ratio of the two partition coefficients and is related to the energy difference between the retention mechanisms for the solutes. Figure

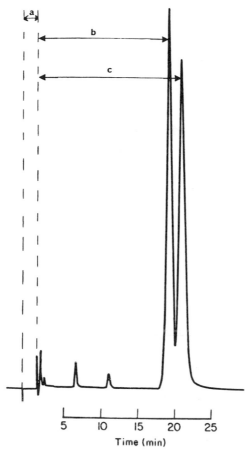

Fig. 1. Typical chromatogram illustrating the terms *capacity ratio* ($\kappa_1 = b/a$; $\kappa_2 = c/a$) and *separability factor* ($\alpha = \kappa_2/\kappa_1$).

1 illustrates these terms. The extent to which two components are actually separated depends both on κ_1 and κ_2 and on the efficiency of the column. Efficiency is a band-shape consideration and is determined by a number of factors, including column-packing skill, particle size, sample size, and flow rate. For a given column the factor most important for the separation of two solutes is the magnitude of α, separations becoming easier as α increases. Because one desires total separation of stereoisomers for both analytical and preparative resolutions, particular attention will be paid to the magnitude of α for the stereoisomers under discussion. Highly efficient analytical HPLC systems will afford reasonably good separations of two solutes having an α of 1.04 or greater, although recycling techniques may have to be employed at the lower end of this

scale. Because MPLC systems typically employ larger particle-sized packings than do HPLC systems, preparative MPLC columns offer reduced efficiency and require α values of perhaps 1.2 or greater for effective resolution of sizable quantities of material.

II. General Considerations

Although it has long been appreciated that, in principle, diastereomers can be separated chromatographically, this approach to the solution of stereochemical problems is only now being used with much frequency. Until a few years ago such separations were sought empirically. It is currently realized that broad classes of diastereomeric derivatives exhibit chromatographic separability; thus, a large number of indirect resolutions may be so effected on both analytical and preparative scales.

For analytical applications the CDA chosen should be of essentially total enantiomeric purity and react nonselectively with both substrate enantiomers to afford stable diastereomers of sufficient separability for the instrumentation at hand. The cost of the reagent is not especially important owing to the small quantities required. Because the indirect resolution method is not an absolute method, incomplete enantiomeric purity of the CDA or differential extents of derivatization of the substrate enantiomers can adversely influence analytical conclusions of substrate enantiomeric purity. These conclusions (drawn from the relative areas of the chromatographic bands stemming from the diastereomeric derivatives) can also be affected if the diastereomers give nonidentical detector responses.

For preparative applications, CDAs should afford diastereomers from which the substrate enantiomers can ultimately be retrieved. Moreover, the chromatographic separability of the diastereomeric derivatives must be greater (owing to reduced column efficiency), and the cost of CDAs becomes a consideration. Differential extents of reaction with the enantiomers and differential detectability of the diastereomers are not serious problems, although it is hoped that overall yields during derivatization and dederivatization will be high. Because an enantiomer retrieved from a chromatographically isolated diastereomer derived from a CDA of incomplete enantiomeric purity will be similarly impure, it is desirable that the CDA be of high enantiomeric purity. However, this requirement is tempered by the realization that recrystallization can raise the enantiomeric purity of an enriched compound. It might be noted that serendipitous crystallization of one diastereomer can simplify preparative work by reducing the amount of material to be chromatographed.

Direct chromatographic resolutions are free of a number of the restraints placed on indirect methods. One major difference is that direct methods are absolute in the sense that no external standard of enantiomeric purity is needed to determine the enantiomeric purity of a substrate with accuracy. Chromatography, like NMR, gives a weighted time-averaged view of dynamic processes. Hence, a less than enantiomerically pure CSP or chiral mobile-phase additive (CMPA) affects the position but not the relative size of the two bands stemming from the solute enantiomers. Reduction of the enantiomeric purity of a CSP or CMPA diminishes α by increasing κ_1 and diminishing κ_2 until, at the racemic limit, the two values become identical. If one uses a column packed with a CSP of but 80% enantiomeric purity, the column will still function effectively, although the observed $\alpha - 1$ values will be $\sim$ 80% of the maximum attainable. From the preparative standpoint, direct resolutions may afford higher overall yields because derivatization and dederivatization may be unnecessary. However, derivatization is sometimes employed either to enhance detectability or to provide functionality necessary for the separation process. Since the derivatizing agents employed are achiral, no significant possibility exists for selective derivatization of one enantiomer, and incomplete derivatization is of little consequence for analytical applications.

Offsetting the overall advantage of the direct method is the fact that only a few types of chiral chromatographic columns or CMPAs are commercially available at present. However, once a chiral column is in hand, it can be used a great many times, thereby compensating for the initial expense or difficulty in obtaining the column. The CMPA method requires an ongoing source (or recovery) of the chiral additive. This problem will be alleviated with the advent of microbore analytical columns (reduced mobile-phase consumption) but will remain serious for preparative separations. Moreover, the CMPA must ultimately be separated from each enantiomer after a preparative resolution.

As mentioned earlier, enantiomeric purity is determined from the relative areas of the chromatographic bands stemming from the stereoisomers under comparison. In the CDA and CMPA methods the bands stem from diastereomers that need not give identical detector responses. In the CSP approach one detects enantiomers that will give identical responses.

Absolute configurations can be determined by any chromatographic method if a configurationally known sample is available for comparison. Failing this, absolute configurations can be assigned on the basis of elution order, provided that one knows the absolute configuration of the chiral agent and the mechanism of chromatographic separation for the relevant stereoisomers. An additional advantage of the indirect method is that the covalent diastereomers may differ in their NMR parameters and it may be possible to correlate tnese differences with relative and hence absolute configuration. Short-lived diastereomeric solvates can also differ in their NMR parameters, and these differences can also be correlated to relative/absolute stereochemistry. However, few such correlations have yet been made for the types of chiral agents shown to be useful as CSPs or CMPAs.

III. Indirect Resolutions

The first attempts to systematize the chromatographic separation of diastereomers were made by Westly and co-workers (1968). Westly's gas chromatographic separation of diastereomeric esters is now largely of historical interest as a foundation on which subsequent liquid chromatographers have built.

A. Diastereomeric Amides, Carbamates, and Ureas

1. SCOPE AND CHEMISTRY

Helmchen and co-workers (Helmchen et al., 1972, 1977, 1979a,b; Helmchen and Nill, 1979) have focused on the separation of diastereomeric amides (**1**) derived from acids (or lactones) and amines. One of the reaction partners is used as the CDA and must be available as a single enantiomer. Helmchen has separated a wide variety of amide diastereomers on silica gel, the reported α values (Table I) being sufficiently large for straightforward preparative resolutions. Subsequent retrieval of the chiral acid can be complicated by racemization during hydrolysis. Helmchen typically uses acid-promoted hydrolysis and reports that amides containing an appropriately situated hydroxyl group will hydrolyze under mild conditions without racemization of the acid component.

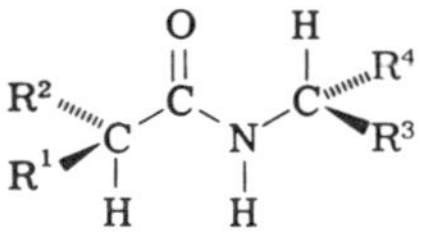

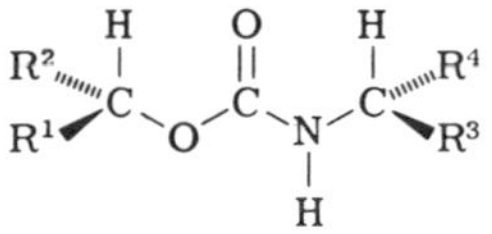

A series of diastereomeric carbamates has been examined by Pirkle and Hauske (1977a), Pirkle and Boeder (1978a), Pirkle and Rinaldi (1978), and Pirkle and Adams (1979). These carbamates (**2**) are derived either from alcohols and isocyanates or from chloroformates and amines. In each case one of the components is the CDA and must be available as a single enantiomer. Table II provides α values for a number of carbamate diastereomers, these values generally being adequate for facile preparative separation on large silica or alumina columns. After separation, enantiomerically pure alcohol can be retrieved from the carbamate by the use of trichlorosilane (Pirkle and Hauske, 1977b), a reagent

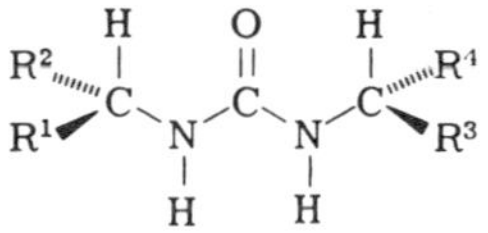

TABLE I
Chromatographic Separation of Diastereomeric Amides

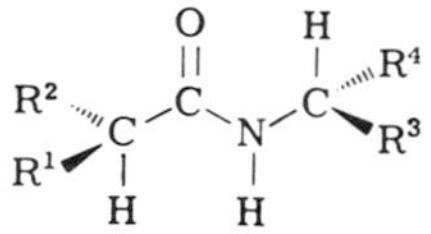

$$R^2\text{---}\underset{R^1}{\overset{H}{C}}\text{---}\overset{\overset{O}{\|}}{C}\text{---}\underset{H}{N}\text{---}\underset{H}{\overset{H}{C}}\overset{R^4}{\underset{R^3}{}}$$

1

R^1	R^2	R^3	R^4	α^a
Et	Ph	Me	α-Naphthyl	1.80
Et	Ph	Me	Ph	1.51
Me	Ph	Me	Ph	1.81
Me	CH$_2$Ph	Me	Ph	1.68
Me	nBu	Me	p-NO$_2$C$_6$H$_4$	2.21
Me	Ph	Ch$_2$OH	Ph	2.56
Me	CH$_2$Ph	Ch$_2$OH	Ph	2.56
Me	n-Hexadecyl	Ch$_2$OH	Ph	2.08
CH$_2$OH	Ph	Me	Ph	2.81
(CH$_2$)$_2$OH	Ph	Me	Ph	2.83
(CH$_2$)$_3$OH	Ph	Me	Ph	2.13
(CH$_2$)$_2$OH	Ph	Ch$_2$OH	Ph	5.77

^{a}The diastereomer with the relative configuration shown is eluted last.

TABLE II
Chromatographic Separation of Diastereomeric Carbamates

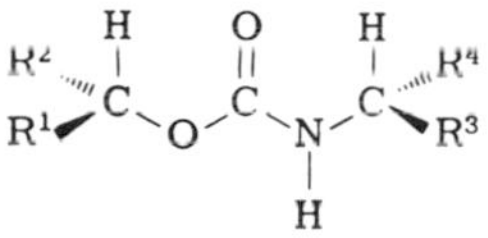

$$R^2\text{---}\underset{R^1}{\overset{H}{C}}\text{---}O\text{---}\overset{\overset{O}{\|}}{C}\text{---}\underset{H}{N}\text{---}\underset{H}{\overset{H}{C}}\overset{R^4}{\underset{R^3}{}}$$

$R^3 = \alpha$-naphthyl; $R^4 = CH_3$

2

R^1	R^2	α^a	R^1	R^2	α^a
CF$_3$	Ph	1.58	Me	(CH$_2$)$_3$CN	2.8
CF$_3$	p-FC$_6$H$_4$	2.04	Me	(CH$_2$)$_4$CN	1.4
CF$_3$	α-Naphthyl	1.56	n-Octyl	(CH$_2$)$_2$CN	2.0
9-Anthryl	CF$_3$	1.40	C$\equiv$C(CH$_2$)$_8$CH$_3$	(CH$_2$)$_3$CN	1.8
n-C$_3$F$_7$	Ph	2.12	CH$_2$SC$_6$H$_5$	Me	1.25
n-C$_3$F$_7$	α-Naphthyl	2.12	CH$_2$SC$_6$H$_5$	Et	1.16
Ph	Me	1.30	Ph	CH$_2$SC$_6$H$_5$	1.30
α-Naphthyl	Me	1.20	CH$_2$SCH$_2$CH$_3$	n-Octyl	1.17
Et	Ph	1.22	C$\equiv$CH	Me	1.13
Et	α-Naphthyl	1.25	C$\equiv$CH	Et	1.21
tBu	α-Naphthyl	1.31	C$\equiv$CH	nBu	1.27
tBu	Ph	1.30	C$\equiv$CH	n-Hexyl	1.43

^{a}The diastereomer with the relative configuration shown above was eluted last.

that functions under mild conditions in a variety of solvents and that tolerates the presence of a number of other functional groups in the carbamate.

Although a number of diastereomeric ureas are known to be chromatographically separable, no systematic study of ureas such as **3** has been reported. Such ureas are usually rather insoluble, are quite polar, and, if hydrolyzed, would give two amine components. Unsymmetric acylureas such as **4** and **5** have been studied and resolve quite well (Table III) in a systematic manner

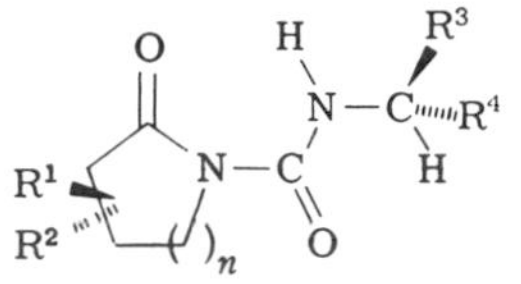

4 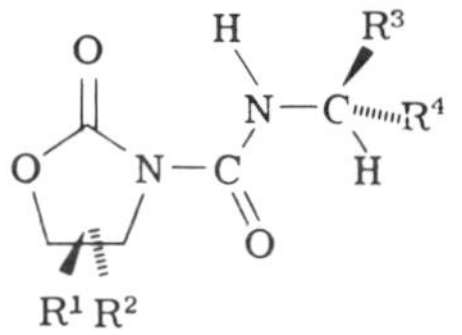   **5**

(Robertson, 1981; Pirkle and Simmons, 1982). These compounds, derived from the action of isocyanates on lactams (**6**) and oxazolidinones (**7**), can be dederivatized smoothly. This method offers great promise for the resolution of a number of nitrogen heterocycles. Alternatively, oxazolidinone CDAs such as **8** can be used to resolve amines.

TABLE III
Chromatographic Separation
of Diastereomeric Urea Derivatives 4 and 5

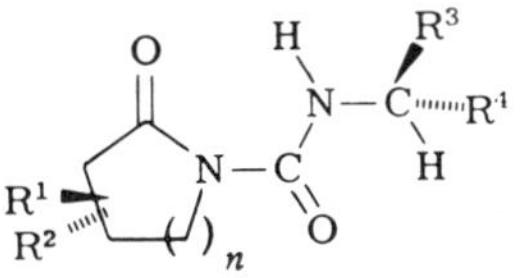

4 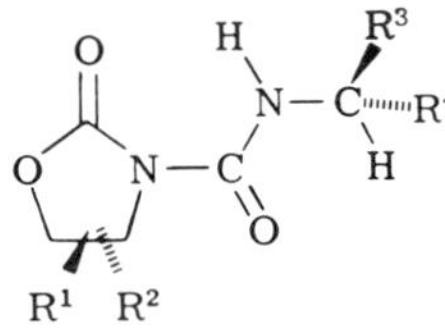   **5**

$R^2 = H$; $R^3 = \alpha$-naphthyl; $R^4 = Me$

Derivative	R^1	α^a
4a	5-(C_6H_5)	2.29
4b	5-(p-$CH_3C_6H_4$)	2.33
4c	5-(p-FC_6H_4)	2.47
4d	4-(C_6H_5)	2.04
4e	3-(C_6H_5)	1.84
5a	4-(C_6H_5)	2.05
5b	5-(C_6H_5)	2.32
5c	4-(α-Naph)	2.37
5d	5-(α-Naph)	2.61

[a]The diastereomer with the relative configuration shown above is eluted last.

6   **7**   **8**

2. MECHANISM OF CHROMATOGRAPHIC SEPARATIONS

Two substances separate chromatographically owing to a blend of differential solvation by the mobile phase and differential adsorption by the stationary phase. Nuclear magnetic resonance and infrared studies of the aforementioned amides, carbamates, and ureas strongly indicate that these compounds have more or less semirigid planar backbones composed of all the atoms in **1–5** *not* indicated to project from the page. Extensive population of the conformations shown in **1–5** stems from combinations of hydrogen bonding, dipolar repulsion, steric, and carbinyl hydrogen bonding effects. Owing to multiple-site binding of the polar backbone to the adsorbent, the backbone lies more or less flat on the adsorbent with the two substituents projecting from the "bottom" face of the backbone directed toward the adsorbent. The "top-face" substituents project away from the adsorbent. The "bottom-face" substituents provide important interactions with the adsorbent, which may be either repulsive or attractive. The relative disposition of the substituents (e.g., whether R^1 is syn or anti to R^3 in the depicted conformations) coupled with the extent and sense of their interaction with the adsorbent determines the elution order of the two diastereomers. Differential solvation seems unimportant. When R^1 is more "repulsive" toward the adsorbent than R^2, and R^3 is more "repulsive" than R^4, the diastereomer having R^1 syn to R^3 will be more tightly adsorbed (from the unhindered face that bears R^2 and R^4) than will its "anti" counterpart (in which both faces are hindered). If R^1 is more "attractive" than R^2, and R^3 is more "repulsive" than R^4, the diastereomer having R^1 syn to R^4 will be the more strongly retained. Alternatively, if R^1 is more "attractive" than R^2, and R^3 is more attractive than R^4, the diastereomer having R^1 syn to R^3 will be the more strongly retained. By using such simple concepts and ranking the interactive capacity of substituents, one can correlate stereochemistry and elution order remarkably well. The actual magnitude of interaction of a given substituent with the adsorbent depends on the adsorbent, other substituents present, and the type and rigidity of the backbone itself. Although no serious attempts at quantifiation have been made, repulsive interactions toward silica and alumina can be ranked roughly as H < methyl < phenyl ≈ ethyl < propyl < *tert*-butyl < trifluoromethyl < α-naphthyl < 9-anthryl ≈ pentafluoroethyl < heptafluoropropyl. Size and hydrophobicity are both relevant; incorporation of polar functionality (hydroxyl, carbalkoxy, cyano) leads to attractive rather than repulsive interactions with silica or alumina.

The drawings shown in Tables I–III represent the stereochemistry of the more strongly retained diastereomer. The NMR spectral differences between diastereomers can sometimes be used to check stereochemical inferences drawn from elution orders. In general, an aryl substituent will more heavily shield a syn than an anti substituent.

One will note that the chiral acyclic subunits of 1–5 bear a single hydrogen on the chiral center. The presence of this hydrogen confers a degree of conformational control (essential to separation), which may not be present if some substituent other than hydrogen is present. However, if conformational rigidity is still present, separation of the diastereomers might still occur for essentially the reasons just advanced. It is also important that the chiral amine subunits of 1–5 be primary so that the (Z)-amide rotamer is preferentially populated. Diastereomers derived from secondary amines have been observed to separate but with diminished α values.

3. APPLICATIONS

Subject to the limitations discussed in Section II, the chromatographic separation of diastereomeric amides and carbamates can be used to assay enantiomeric purities and assign absolute configurations for a variety of alcohols, acids, and amines. In applying this method to chiral acids, diastereomeric amides derived from 1-(phenyl)ethylamine (e.g., see Helmchen and Strubert, 1974; McKay et al., 1979) or 1-(α-naphthyl)ethylamine (e.g., see Eberhardt et al., 1974) are typically used. Valentine et al. (1976), Scott et al. (1976), and Bergot et al. (1978) have applied this method to a series of isoprenoid and terpenoid acids using (R)-1-(p-nitrophenyl)ethylamine or 1-(α-naphthyl)ethylamine as CDAs (Table IV). It is important to note that sizable α values are observed even though the chiral center of the acid is separated by a methylene unit from the amide functionality.

The chromatographic separation of diastereomeric hydroxyamides has been utilized to obtain optically active acids and amines on a preparative scale. Helmchen et al. (1979a) have shown that hydroxyamides exhibit very favorable chromatographic properties (see α values in Table I) and are hydrolyzed under mild nonracemizing conditions, overall resolution yields typically being 80–90%. (R)-Phenylglycinol (9a) is suggested as a CDA for the resolution of acids, whereas 1-(phenyl)ethylamine (9b) is suggested for the resolution of lactones such as 10a and 10b. Conversely, amines can also be resolved by using lactone 10c as a CDA. The observed elution order for all hydroxyamides examined by Helmchen are those expected on the basis of the described chromatographic model. Hence, relative/absolute configurations of the resolved components can be assigned.

Ade and co-workers (1980) have utilized the chromatographic separability of

TABLE IV
Resolution of Chiral Isoprenoid Acids
via the Chromatographic Separation
of a Diastereomeric Amide Derivative

Acid component	Amine component[a]	α
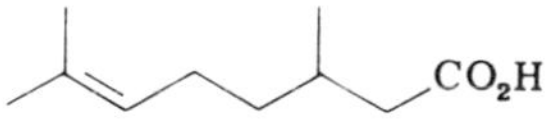	A B	1.22 1.21
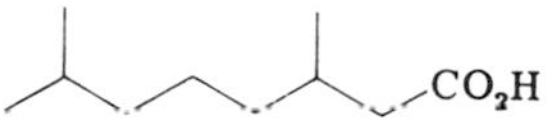	A B	1.21 1.24
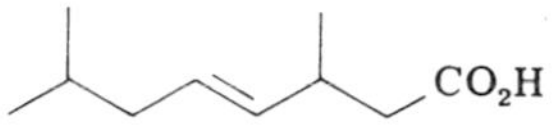	A	1.47
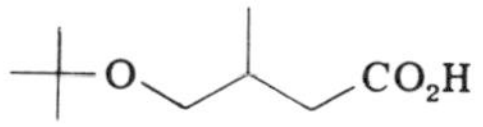	A	1.03
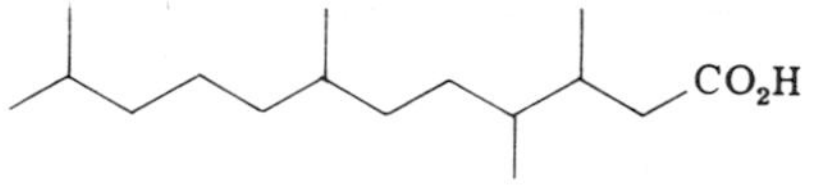	A	1.05
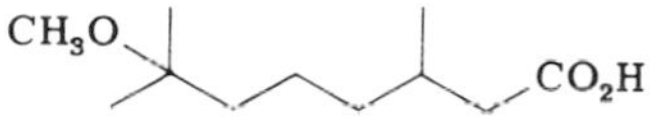	B	1.09

[a] A, 1-(p-Nitrophenyl)ethylamine; B, 1-(α-naphthyl)ethylamine.

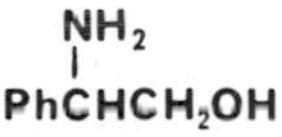

| 9a | 9b | 10a | 10b | 10c |

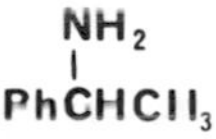

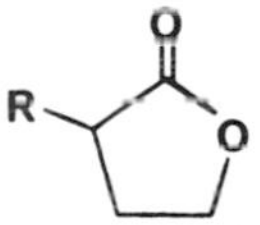

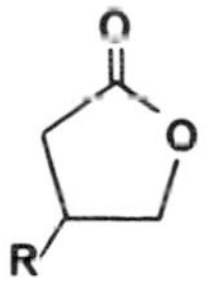

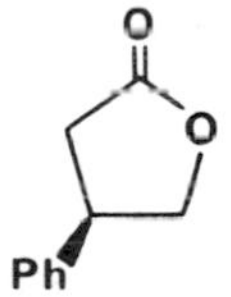

diastereomeric hydroxyamides in the synthesis of the optically active hydrocarbon **11**, the pheromone of the tsetse fly. In this synthesis, diastereomeric amides (**12**) were separated by chromatography and then hydrolyzed. Due to the relatively large α value, the diastereomeric amides were readily resolved in 3-g lots on a MPLC system (Fig. 2). The optically active acid was subsequently converted to the pheromone by standard methods.

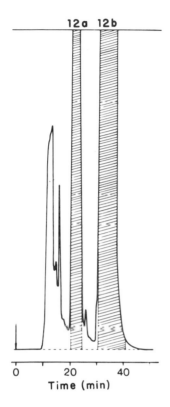

12a   12b

0   20   40
Time (min)

Fig. 2. Preparative LC separation of the diastereomeric amides **12a** and **12b**. Column: 45 × 4 cm, 270 g silica gel Merck LiChroprep, 15–25 μm [9000 theoretical plates; standard test conditions: naphthalene, hexane–ethyl acetate, 95:5 (50 ml/min)]; eluant: chloroform–ethyl acetate, 6/4 (30 ml/min); detector: 254 nm; substance load: 3 g raw reaction product dissolved in 60 ml eluant. From Ade *et al.* (1980). Copyright 1980, Pergamon Press, Ltd.

$$CH_3(CH_2)_{15}\overset{\overset{\displaystyle CH_3}{|}}{C}HCH_2CH_2CH_2CH(CH_2)_{15}CH_3$$

11

$$CH_3(CH_2)_{15}\overset{\overset{\displaystyle CH_3}{|}}{C}HCONH\overset{\overset{\displaystyle Ph}{|}}{C}HCH_2OH$$

12

Enders and Lotter (1981) have demonstrated that diastereomeric α-hydroxyamides (**14**) can be readily separated with α values ranging between 1.5 and 2.0. The resolved α-hydroxyamides are versatile intermediates because they can be converted to optically active α-hydroxy ketones (**15**), diols (**16**), or α-hydroxy acids (**17**) (Scheme 1). A chromatographic model correlating elution orders and α values with stereochemistry and structure was not proposed.

A number of amino acids have been indirectly resolved by HPLC. Generally, the CDA is used to derivatize the amine group, the carboxyl group being previously converted to an ester. Of the derivatives examined, amides (Goto *et al.*, 1978) and thioureas (Nambara *et al.*, 1978) such as **18–20** seem to be the most suitable. Although the amide derivatives are the most readily separated, the isothiocyanate CDAs are more readily accessible, and the thiourea chromophore facilitates ultraviolet detection.

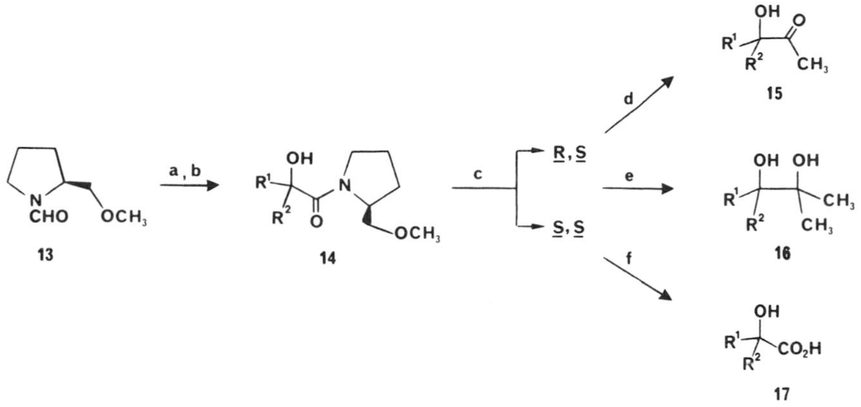

Scheme 1. (a) LiTMP, −100°C. (b) R^1COR^2. (c) Chromatography. (d) 1 equivalent MeLi. (e) 2 equivalents MeLi. (f) Concentrated HCl.

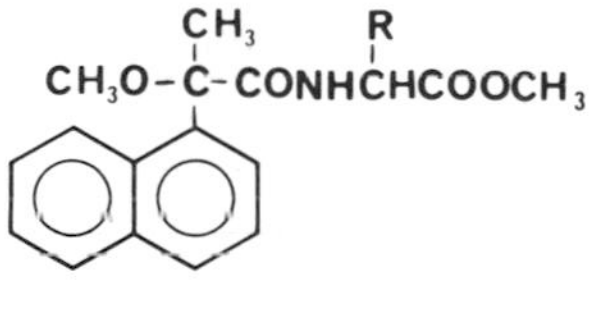

18

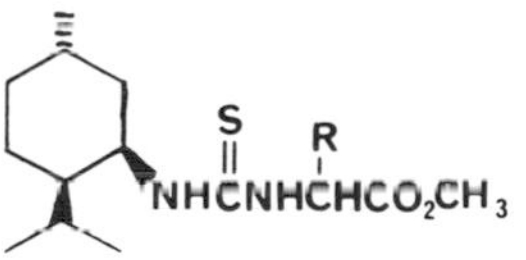

19

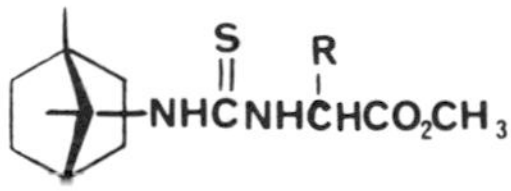

20

Chromatographic separability of diastereomeric carbamates has proved to be a useful and general method for preparatively resolving secondary alcohols. Typically, carbamates derived from either 1-(α-naphthyl)ethyl isocyanate (**21a**) or 1-phenylethyl isocyanate (**21b**) are used. The use of **21a** usually leads to higher α values, but **21b** is much less expensive. Pirkle and Hoekstra (1974), Pirkle and Hauske (1977a), and Finn (1982) have used this method to resolve a series of trifluoromethylcarbinols (**22a–22d**), which are of value as chiral solvating agents and/or precursors of CSPs (see Section IV). Figure 3 shows the chromatogram for the automated repetitive resolution of 1.0-g samples of the carbamate derived from **22d** and CDA **21a**. By the use of larger scale operations, hundred-gram quantities of resolved carbinols have been obtained.

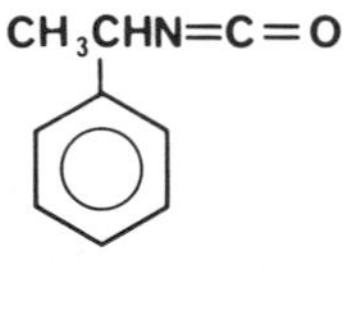

21b

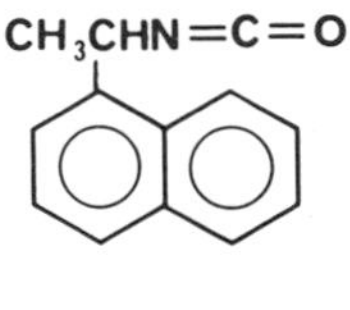

21a

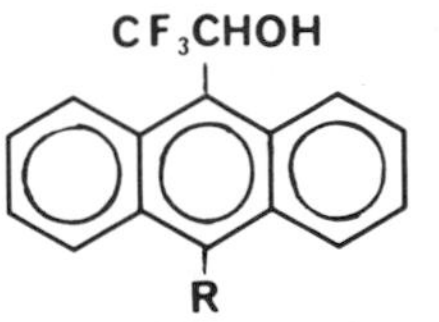

22a R = H
b R = CH_3
c R = OCH_2CH_2Br

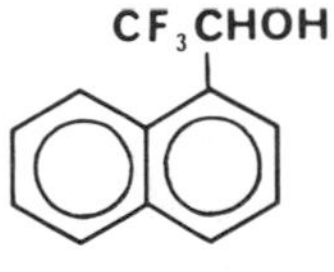

22d

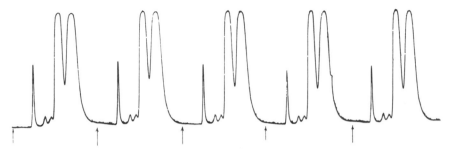

Fig. 3. Automated repetitive chromatographic separation of the diastereomeric carba-mates derived from **21a** and **22d** on acidic alumina with benzene. The separability factor α is 1.37. The *R,R*-diastereomer is the first of the two major bands; minor absorptions are caused by impurities. Sample injections of 1 g (arrows) occur every 3 h. Because of satura-tion of the 280-nm detector, the extent of peak overlap appears to be greater than is actually the case. From Pirkle and Hoekstra (1974). Copyright 1974 American Chemical Society.

Mori and Sasaki (1980) resolved alcohol **23**, a key intermediate in the syn-thesis of the pheromone lineatum, by chromatographically separating the di-astereomeric carbamates derived from CDA **21a**. Schmid and co-workers (1979) utilized **21b** to resolve alcohol **24**, an intermediate in their monensin synthesis.

Diastereomeric carbamates derived from tertiary alcohols can also be sepa-rated chromatographically, as demonstrated by Corey and co-workers (1978) in the synthesis of gibberellic acid, in which the resolution of alcohol **25** was accomplished using the carbamates derived from **21b**.

23	**24**	**25**

The advantage of indirect carbamate resolutions is that the selectivity of the derivatization and dederivatization reactions permits the presence of additional functionality. Diastereomeric carbamates derived from cyano alcohols (**26**) are valuable for the broad-spectrum synthesis of optically active lactones (Pirkle and Adams, 1979, 1980a,b). These diastereomers are generally easy to separate (see Table II), and trichlorosilane cleavage and nitrile hydrolysis readily lead to optically active lactones (**27**, Scheme 2). A number of optically active lactonic pheromones have been so prepared.

Optically active epoxides (**29**) can be obtained by carbamate resolution of β-hydroxysulfides (**28**), S-alkylation, and treatment with base (Pirkle and Rinaldi,

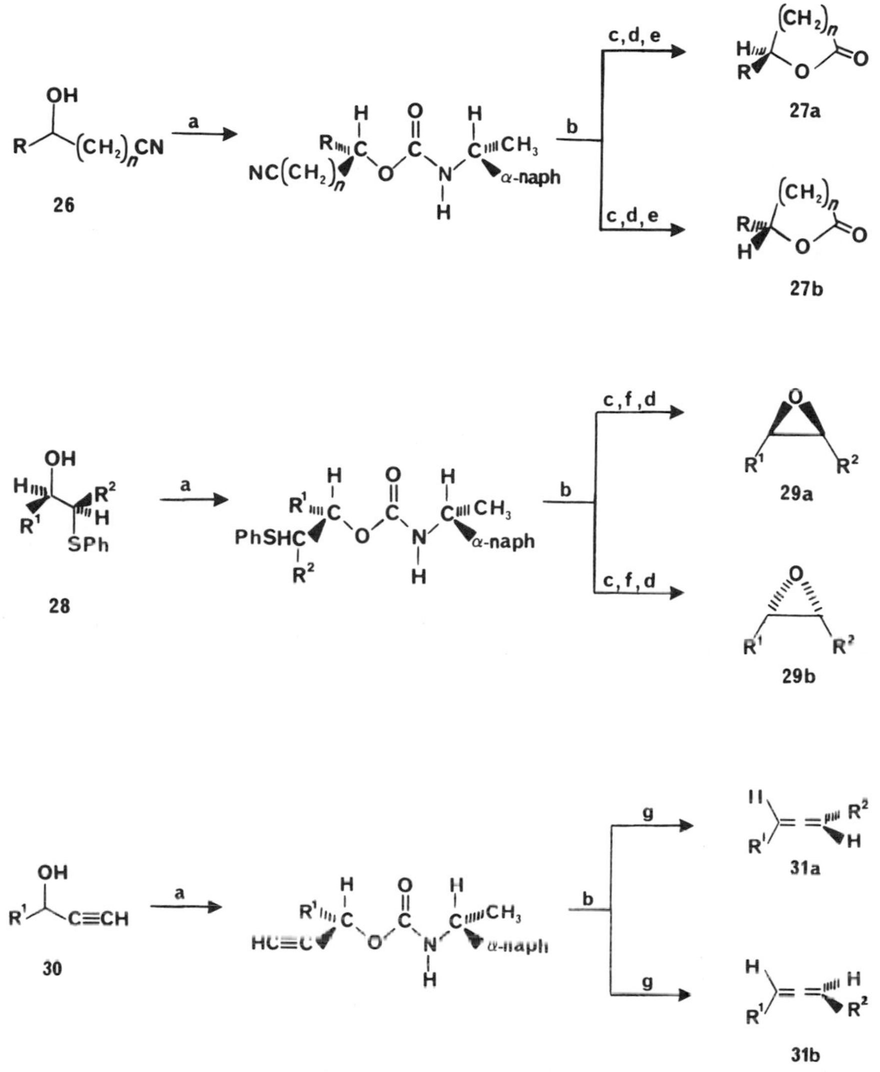

Scheme 2. (a) 21a. (b) Chromatography. (c) HSiCl$_3$. (d) HO$^-$. (e) Lactonization. (f) Me$_3$OBF$_4$. (g) R$_2^2$CuLi.

1978) (Scheme 2). This route was used in the synthesis of (+)-disparlure, the sex attractant of the gypsy moth (Pirkle and Rinaldi, 1979). Pirkle and Boeder (1978a) similarly resolved propargylic alcohols (30), some of which were used to prepare optically active allenes (31), including the sex pheromone of the male dried bean beetle (Pirkle and Boeder, 1978b).

B. Resolutions Using Miscellaneous Types of Diastereomers

One might surmise that some of the same considerations that influence the separability of diastereomeric amides, carbamates, and ureas would be relevant to the separation of the diastereomers of the various other classes of derivatives for which such separations have been noted. However, there has been little systematic study of such derivatives, and little more than a survey of these separations can be offered.

1. ESTERS AND CARBONATES

In the author's experience the diastereomers of esters and carbonates separate much less well on silica or alumina than do those of structurally similar amides and carbamates. Possibly, esters have less tendency to "lie flat" on the adsorbent because, lacking amide hydrogens capable of hydrogen bonding, the esters are principally "adsorbed" at the carbonyl oxygen. For example (Robertson, 1981), the diastereomers of carbamates **32a** and **32b** are readily resolved on

32a Ar = α-naph
b Ar = Ph

33a Ar = α-naph
b Ar = Ph

silica with α values of 1.36 and 1.67, respectively, whereas the diastereomers of the corresponding carbonates (**33a** and **33b**) exhibit separability factors of 1.19 and 1.25, respectively. Diastereomeric esters that possess an additional polar site may separate somewhat better (on polar adsorbents), possibly because this second polar site tends to cause them to "lie flat" on the adsorbent or otherwise confer conformational rigidity. Leitch and co-workers (1968) demonstrated such separations for lactates (**34**) and mandelates (**35**) (Table V). Corey et al. (1979) utilized the chromatographic separability of mandelates (**36**) in the synthesis of

36

TABLE V
Chromatographic Separation of Diastereomeric Esters

Structure	α	Structure	α
OH CH$_3$ | | CH$_3$CHCO$_2$CH(CH$_2$)$_2$CH$_3$ **34a**	1.06	OH CH$_3$ | | PhCHCO$_2$CH$_2$ CHCH$_2$CH$_3$ **35a**	NR[a]
		OH CH$_3$ | | PhCHCO$_2$CH(CH$_2$)$_2$CH$_3$ **35b**	NR[a]
OH CH$_3$ | | CH$_3$CHCO$_2$CHCH(CH$_3$)$_2$ **34b**	1.11	OH CH$_3$ | | PhCHCO$_2$CHCH(CH$_3$)$_2$ **35c**	1.14

[a]NR, Not resolved.

erythronolide A. Mösher's NMR CDA, α-methoxy-α-trifluoromethylphenyl-acetyl chloride, has been used in the preparative resolution of an abscissic acid intermediate (Koreeda et al., 1973) and juvenile hormone intermediate (Imai and Marumo, 1976). Esters derived from camphoric acid (Kuritani et al., 1977) or from menthol (Boyd et al., 1978) have been used in indirect resolutions.

An attractive feature of indirect ester resolutions is that the desired alcohol (or acid) component is readily retrieved by hydrolysis after preparative separation. Because alcohols can be retrieved from carbamates under even milder conditions (trichlorosilane cleavage) and the corresponding carbamate diastereomers offer increased chromatographic separability, the use of an indirect carbamate resolution will usually be preferred for alcohol resolutions.

2. DIASTEREOMERIC OXAZOLINES

Meyers and co-workers (Meyers et al., 1979; Meyers and Slade, 1980) have reported the preparative separation of the diastereomers of oxazolines 37 and 38. After separation, the diastereomers of oxazoline 37 were converted to enantiomeric α-hydroxy acids (17); diastereomeric oxazolines 38 similarly provided a variety of optically active acids (39). The degree of separability of these diastereomers was not reported.

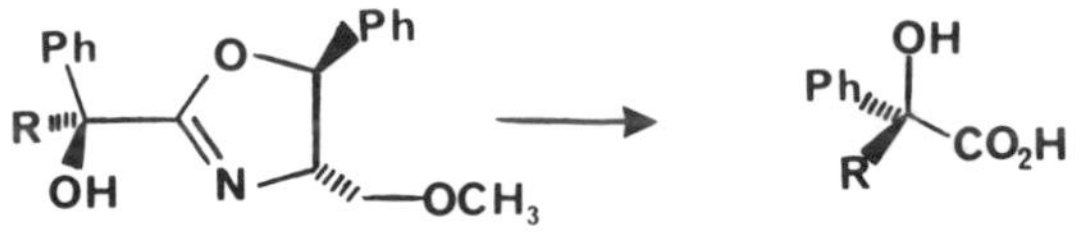

37   17A

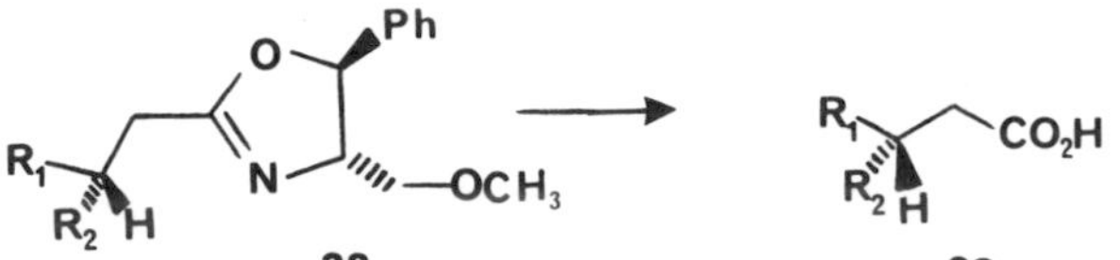

38   39

3. Diastereomers Having an Asymmetric Sulfur Atom

The chromatographic separability of an assortment of diastereomeric sulfoxides, hydroxysulfoxides, sulfilimines, and sulfonamides has been demonstrated. Szokan and co-workers (1980) have examined the chromatographic separability of several sulfoxides (40) and sulfilimines (41) (Table VI). The chromatographic separation of diastereomeric β-hydroxysulfoxides has been exploited in asymmetric synthesis by Farnum *et al.* (1977) in the synthesis of disparlure and by Williams and Phillips (1981) in the synthesis of juvabiols. Diastereomeric sulfonamides (42) can be used in the chromatographic resolutions of chiral amines (Souter, 1976). The principal difficulty encountered in this series is that not many enantiomerically pure sulfoxides are available for use as CDAs.

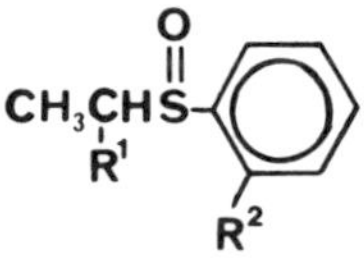

40 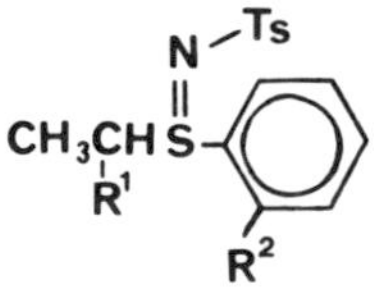   41

42

4. Diastereomers Having an Asymmetric Phosphorus Atom

Guyon *et al.* (1978) have demonstrated that diasteromeric phosphonates (43) are chromatographically separable. After chromatographic resolution, optically active phosphines (44) were prepared from the diastereomeric phosphonates.

TABLE VI
Chromatographic Separation of Diastereomeric Sulfoxides and Sulfilimines

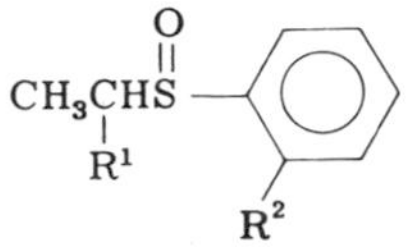

$$
\begin{array}{cc}
\text{40} & \text{41}
\end{array}
$$

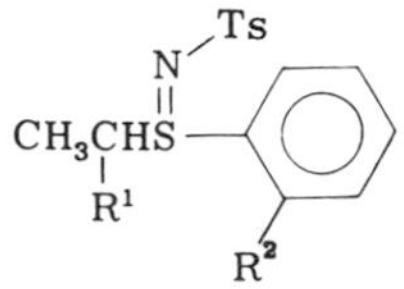

Compound	R^1	R^2	α	Compound	R^1	R^2	α
40a	$n\text{-}C_6H_{13}$	H	1.36	41a	$n\text{-}C_6H_{13}$	H	1.46
40b	$n\text{-}C_2H_5$	H	1.47	41b	$n\text{-}C_2H_5$	H	1.49
40c	$n\text{-}C_6H_{13}$	CO_2H	1.38	41c	$n\text{-}C_6H_{13}$	CO_2H	1.16
40d	$n\text{-}C_2H_5$	CO_2H	1.23	41d	$n\text{-}C_7H_5$	CO_2H	1.24

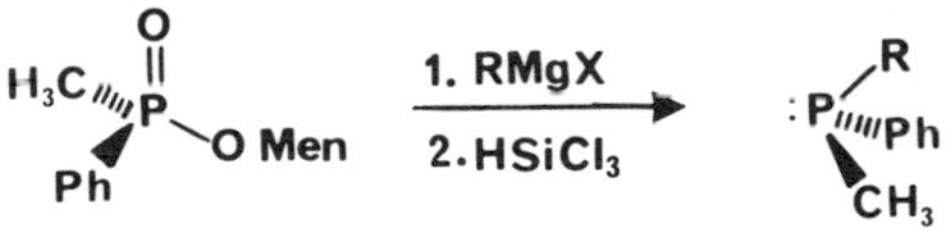

$$
\text{43} \quad \xrightarrow[\text{2.HSiCl}_3]{\text{1. RMgX}} \quad \text{44}
$$

5. KETONE DERIVATIVES

Derivatization of ketone carbonyls with a CDA has provided chromatographically separable diastereomers. Pappo and co-workers (1973) have shown that the chiral oxime 45, a prostaglandin intermediate, is chromatographically separable even though the asymmetric centers are relatively far apart. Boding and Musso (1978) have similarly separated diastereomeric dithioketals (46). This method is limited, however, because enantiomerically pure dithiol CDAs are not widely available.

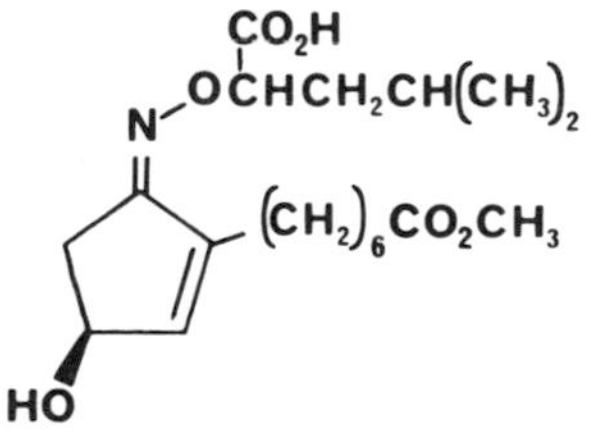

45

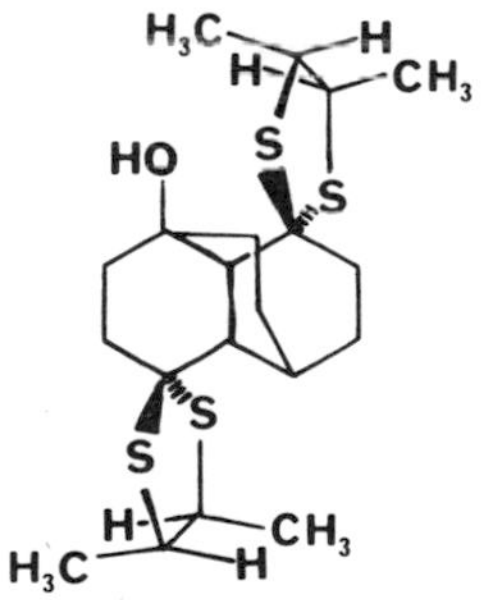

46

IV. Direct Resolutions

It is widely understood that, in principle, enantiomers can be chromatographically separated on a CSP. Alternatively, a CMPA might be used toward
the same end. A review (Davanakov, 1980) mentions over 400 published attempts at such direct resolutions. Only in the past several years, however, has
there been more than an occasional success. With recent advances in chromatographic and synthetic methodologies and in mechanistic concepts, direct
resolution techniques have blossomed and are often the method of choice in the
solution of a host of stereochemical problems.

The use of a CSP is more common than the use of a CMPA and, in our
opinion, has some inherent advantages. Once in hand, a chiral LC column (i.e.,
a column packed with a CSP) can be used thousands of times without special
techniques or apparatus being required. Although the CMPA approach frees the
user of special column requirements, it requires a continuing source of the chiral
additive, the presence of which may complicate detection, or it may require
special detection apparatus. In preparative cases a secondary separation of the
resolved enantiomers from the additive will be required. We hasten to point out
that several very satisfactory analytical applications of the CMPA approach have
appeared. To date, the CMPA approach has been outpaced by the CSP approach
in both scope and understanding of the chiral recognition mechanisms by which
enantiomer separations are achieved. Such understanding is essential if one is to
translate elution order data into assignments of absolute configuration.

A. *Naturally Occurring Chiral Stationary Phases*

Columns packed with readily available naturally occurring chiral solid phases
(e.g., quartz, wool, lactose, and potato starch) have often been used in attempts
at direct resolution. Of such CSPs, microcrystalline triacetylated cellulose has
most often enabled investigators to achieve some degree of success. For example, Hesse and Hagel (1976), Häkli *et al.* (1979), and Mintas *et al.* (1981) have
shown that a variety of racemates containing an aromatic substituent have been
resolved analytically and preparatively on triacetylated cellulose. In these cases
no general understanding of the operative chiral recognition mechanism has been
obtained, and the resolutions have been sought empirically.

B. *Synthetic Chiral Stationary Phases*

We divide synthetic CSPs into two categories: cooperative and independent.
By *cooperative,* we mean that the CSP uses an assemblage of subunits acting in

concert to achieve chiral recognition. Although these subunits may or may not be chiral, the assemblage must be. For example, a polymer having a chiral backbone might be a cooperative CSP. An *independent* CSP is one in which chiral molecules, each capable of chiral recognition, are bound to a support and operate independently in distinguishing between enantiomers. Chiral recognition mechanisms can be most easily understood for independent CSPs, and these processes can be studied in solution using unbound molecules. For example, NMR studies of diastereomeric solvates have been instrumental in the development of some independent CSPs.

1. Cooperative Chiral Stationary Phases

Conceptually, one of the simplest ways to prepare a polymeric CSP is to carry out a polymerization reaction in the presence of a nonreactive chiral "template," which, when later removed, leaves a chiral cavity. Wulff and co-workers (1977) have demonstrated this technique by copolymerization of ethylene dimethylacrylate and methyl methylacrylate in the presence of optically active 4-nitrophenyl α-D-mannopyranoside 2,3,4,6-di-*O*-(4-vinylphenyl borate). Partial resolution of 4-nitrophenyl mannopyranoside was initially possible, but the CSP became progressively less effective due to swelling of the polymer and distortion of the chiral cavity.

Schwanghart *et al.* (1977) have prepared several polyacrylamide (**47**) and polymethylacrylamide (**48**) CSPs and have chromatographed a number of race-

$$\begin{array}{c} -(CH_2CH)_n- \\ | \\ CONHCHR^1 \\ | \\ R^2 \end{array} \qquad \begin{array}{c} CH_3 \\ | \\ -(CH_2C)_n- \\ | \\ CONHCHCH_3 \\ | \\ Ph \end{array}$$

47a $R^1 = CH_3$; $R^2 = Ph$ **48**

  b $R^1 = CH_2Ph$; $R^2 = CO_2C_2H_5$

mates on them. In the majority of cases only partial resolution of the enantiomers is achieved, although enantiomerically pure material can be obtained by the collection of multiple fractions and use of recycling techniques. These resolutions are empirical, no trend being observed that would enable a user to surmise which CSP would be most useful for a given racemate. The polyacrylamide CSP **47a** and the corresponding polymethylacrylamide CSP **48** generally show marked differences in their resolving power and are thought to behave as cooperative CSPs. Some pharmaceutically interesting compounds (e.g., **49–51**) have been preparatively resolved by Blaschke and Markgraf (1980) and Blaschke *et al.* (1980). Figure 4 illustrates one of the better resolutions.

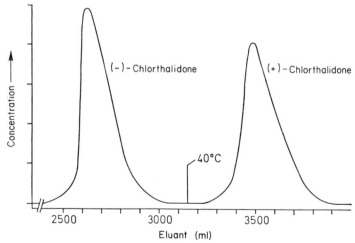

49   **50**   **51**

R = H, CH₃

Yuki *et al.* (1980) have reported the development of an optically active, isotactic polytriphenylmethyl methylacrylate support (**52**), the first polymeric

$$-\!\!\!\left(CH_2CH\right)_{\!n}$$
$$CO_2CPh_3$$

52

phase in which the chirality is due solely to the helicity of the polymer. A chiral anionic catalyst [(−)-spareteine-*n*-butyllithium] is employed to initiate polymerization. In principle, a large quantity of this chiral polymer can be easily prepared, making this CSP potentially useful on a preparative scale. The resolution

Fig. 4. Chromatogrpahic resolution of 530 mg of racemic chlorthalidone (**49**, R = H) on 250 g of **47a**. Column: 76 × 3.2 cm; solvent system: toluene–dioxane (1:1); pressure: 1.2 bars. From Blaschke and Markgraf (1980).

of a number of solutes from different classes was achieved, and a number of the separations were quite impressive. This CSP also appears to operate cooperatively, possibly by intercalation of aryl solute substituents into chiral "channels" in the polymer.

2. INDEPENDENT CHIRAL STATIONARY PHASES

a. Mechanism of Chiral Recognition on an Independent Chiral Stationary Phase

The capacity of a CSP to retain enantiomers differentially is due to molecular interactions between the CSP and the enantiomeric solutes. The degree of chiral recognition manifested in the separability factor α of the enantiomers reflects the stability differences of the diastereomeric complexes stemming from these interactions. Hence, an understanding of the nature of these interactions and how they are influenced by stereochemistry can lead to both a means of relating elution order to absolute configuration and a means of estimating the magnitude of α.

For chiral recognition to occur, the CSP must have a minimum of three simultaneous interactions with at least one of the solute enantiomers, at least one of these interactions being stereochemically dependent. This principle was first stated by Dalgleish (1952) and, in a simple form, is illustrated by Fig. 5. The CSP contains three sites (A, B, and C) capable of simultaneously bonding to a solute having the appropriate complimentary sites (A', B', and C'). Although both solute enantiomers are capable of bonding to the CSP at two sites, only one of them can also undergo the third simultaneous interaction. The diastereomeric complex capable of three simultaneous bonding interactions will be more stable than its "two-bond" counterpart. Had one of the interactions been repulsive rather than attractive, the stability order would have been reversed. The solute enantiomer incorporated into the more stable of the diastereomeric complexes will be the more strongly retained on the column.

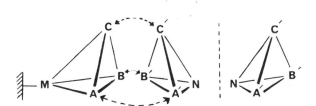

Fig. 5. Three-point interaction model. Only one solute enantiomer possesses the correct chirality to have three simultaneous interactions with the resolving agent.

All types of molecular interactions (e.g., hydrogen bonding, dipole–dipole forces, charge-transfer complexes, steric repulsions, and hydrophobic attractions) would appear to be suitable contributors to chiral recognition. For optimum results three different types of interactions should be involved (to prevent interchange of sites), and spatial orientation of the solute and CSP sites must "match." Both structure and conformational behavior enter into these considerations.

b. Examples of Independent Chiral Stationary Phases

The first step in the synthesis of an effective independent CSP involves the acquisition of a chiral compound containing the three (or more) sites necessary to differentiate between the enantiomers of the solute (or class of solutes) to be resolved. This chiral compound is then bound to a solid support (e.g., polystyrene, silica) in such a way as not to interfere with the chiral recognition process. In general, columns packed with a CSP prepared on a polymeric support seem not to give the high chromatographic efficiency obtainable with silica-based bonded phases. Silica-based CSPs have been prepared by several workers, either by bonding a chiral silane to silica or by a stepwise process whereby the chiral compound is bound to a functionalized silane previously attached to silica (Scheme 3). Silica-bound CSPs have the desirable features of being chemically and dimensionally stable to a variety of solvents and of affording efficient columns.

Baczuk and co-workers (1971) were pioneers in the rational design of synthetic CSPs. These workers used cyanuric chloride to bond L-arginine to Sephadex and obtained a CSP that successfully resolves ($\alpha = 1.6$) the enantiomers of β-3,4-dihydroxyphenylalanine (DOPA). This "rational" CSP incorporates three interactive sites for chiral recognition of the DOPA enantiomers (Fig. 6). The partial resolution of tyrosine on this column and the lack of resolution of phenylalanine are consistent with the three interactions deemed operative.

The direct resolution of chiral helicenes has been accomplished by several groups. Mikes and co-workers (1976) nicely resolved the enantiomers of 10 racemic [5–14]helicenes on 2-(2,4,5,7-tetranitro-9-fluorenylideneaminoxy)propionic amide CSP. Similar resolutions were effected by Lochmüller and Ryall

$$(RO)_3Si(CH_2)_n R^* \ + \ \text{silica gel} \ \longrightarrow \ \equiv Si(CH_2)_n R^*$$

$$53 \qquad\qquad\qquad\qquad\qquad\qquad 54$$

$$(RO)_3Si(CH_2)_n Y \ + \ \text{silica gel} \ \longrightarrow \ \equiv Si(CH_2)_n Y \ \xrightarrow{\overset{*}{R}CH_2X} \ \equiv Si(CH_2)_n Y\, CH_2R^*$$

$$55 \qquad\qquad\qquad\qquad\qquad\qquad 56 \qquad\qquad\qquad\qquad\qquad 57$$

Scheme 3. $R^* = $ chiral functionality; $\equiv$Si symbolizes a silane bound to a solid support.

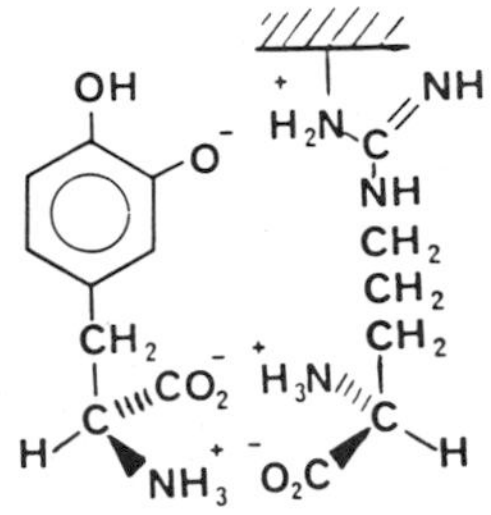

Fig. 6. Three-point interaction mechanism between (S)-DOPA and Baczuk's (S)-arginine CSP.

(1978) using a column packed with an N-(2,4-dinitrophenyl)alanine amide CSP and by Mikes and Boshart (1978) using a column packed with a bi-β-naphthol diphosphate amide CSP. The chiral recognition in these examples is due to a combination of π–π charge transfer and steric interactions.

Several CSPs have been devised to resolve α-amino acids. Sousa *et al.* (1978) and Dotsevi *et al.* (1979) resolved the salts of some α-amino esters on chiral crown ether CSPs; impressively large α values were obtained, although column efficiency was low. Hara and Dobashi (1979a,b) effected the resolution of N-acylated amino acids on an amino acid-derived CSP, small α values (typically ranging between 1.05 and 1.38) being offset by high column efficiency. Stewart and Doherty (1973) resolved DL-typtophan on a bovine serum albumin succinoylaminoethyl-Sepharose CSP.

i. FLUORO ALCOHOL CHIRAL STATIONARY PHASES. Chiral fluoro alcohols such as **22a** have proved to be useful as chiral solvating agents for NMR determinations of absolute configuration and enantiomeric purity. NMR studies have shown that this (and similar) fluoro alcohols form two-point, chelate-like solvates with a variety of dibasic solutes; the principal interaction is hydrogen bonding, whereas the secondary interaction is carbinyl hydrogen bonding. The diastereomeric solvates so formed differ in stability only when a simultaneous third interaction is possible for one, but not both solute enantiomers. This condition is met when the solute contains a π-acidic substituent capable of interacting with the π-basic anthryl ring of the fluoro alcohol **22a**. For example, Pirkle and Sikkenga (1975) have shown that the diastereomeric solvates **58a** and **58b**, formed with fluoro alcohol and racemic methyl-2,4-dinitrophenyl sulfoxide, are

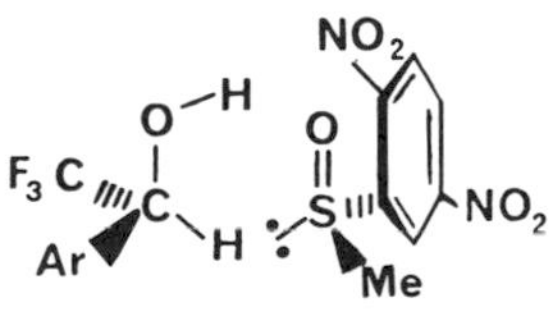

58a

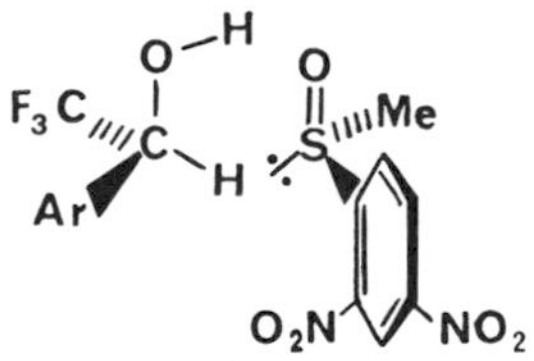

58b

of unequal stability and that this sulfoxide can be resolved on a silica column using the fluoro alcohol as a CMPA (Pirkle and Sikkenga, 1976). Pirkle and House (1979) used 9-(10-bromomethylanthryl)trifluoromethylcarbinol (**59**) to alkylate γ-mercaptopropylsilanized silica to afford CSP **60**. This CSP is capable

F₃C⧹_C⧸OH
   C⁗H

CH₂Br

59

F₃C⧹_C⧸OH
   C⁗H

CH₂S(CH₂)₃Si≡

60

of resolving the enantiomers of essentially all alkyl-2,4-dinitrophenyl sulfoxides (**61**) and a variety of related compounds as well. The elution order of configurationally known sulfoxides is in accord with the model in all instances known to date. Significantly, the enantiomers of a variety of 3,5-dinitrobenzoyl (DNB) derivatives of alcohols (**62**) and amines (**63**) are resolved on this CSP, provided

Ar—S—R (O)

61

O₂N—⬡—C(=O)—O—C(H)(B)(R)

NO₂   **62**

O₂N—⬡—C(=O)—N(H)—C(H)(B)(R)

NO₂   **63**

that these compounds possess the additional basic site required by the chiral recognition model. In these cases chiral recognition can be understood by examination of the more stable diastereomeric complex illustrated by Fig. 7. Because of carbinyl hydrogen bonding, amides (and esters) preferentially populate the conformation shown in which the methine hydrogen lies in nearly the same plane

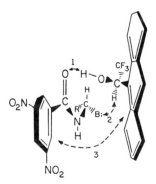

Fig. 7. Three-point interaction possible for the more stable CSP **60**–DNB derivative diastereomeric solvate.

as the amide (ester) carbonyl. For the solute enantiomer shown, three simultaneous bonding interactions are possible with the CSP. The hydroxyl of the fluoro alcohol hydrogen bonds to either the carbonyl oxygen of the DNB group or the alternate basic site, B, depending on which of these two sites is the most basic. The carbinyl hydrogen of the CSP interacts at the remaining basic site. The final interaction is $\pi-\pi$ bonding between the anthryl and the DNB groups. Elution orders of configurationally known solutes support this model.

A representative sampling of the resolutions obtained on fluoro alcohol CSP **60** are compiled in Table VII. Resolutions within a structural class are quite general, although the magnitude of α is somewhat dependent on the structure of the solute. Elution orders are consistent within the class and in accord with the chiral recognition model. An improved fluoro alcohol CSP (**64**) has since been prepared by Finn (1982) (Table VII).

TABLE VII
Direct Resolution of Enantiomers
on Fluoro Alcohol CSPs **60** and **64**

Compound	Ar	R	α on CSP **60**	α on CSP **64**
61a	2,4-Dinotrobenzyl	n-Dodecyl	1.38	1.39
61b	2,4-Dinitrophenyl	n Dodecyl	1.26	1.24
	R	B		
62a	Me	Ph	1.04	1.09
62b	Me	α-Naphthyl	1.08	1.14
62c	Ph	CO_2Me	—	1.29
63a	Me	Ph	1.19	1.28
63b	Et	Ph	1.29	1.35
63c	Me	p-Biphenyl	1.17	1.25
63d	iPr	$CONH$-n-C_4H_9	1.33	1.42
63e	CH_2Ph	$CONH$-n-C_4H_9	1.52	1.67
63f	iBu	$CONH$-n-C_4H_9	1.65	1.87
63g	Ph	$CONH$-n-C_4H_9	1.78	1.79
63h	Ph	CO_2Me	1.19	1.26
63i	Me	CO_2Me	1.08	1.08
63j	Ph	$PO(OEt)_2$	1.38	1.91
63k	iPr	$PO(OEt)_2$	—	1.35
63l	Me	CO_2^-	1.32	1.40
63m	iPr	CO_2^-	1.34	1.36
63n	iBu	CO_2^-	1.35	1.42
63o	Ph	CO_2^-	1.76	1.73
63p	CH_2Ph	CO_2^-	1.39	1.40
63q	sBU	CO_2^-	1.49	1.50

F₃ C   OH
      C""H

ÓCH₂CH₂S(CH₂)₃Si≡

64

ii. AMINO ACID–DINITROBENZOYL-DERIVED CHIRAL STATIONARY PHASES.
The diastereomeric interactions that allow a column derived from chiral A to
resolve racemic B also allows a column derived from chiral B to resolve racemic
A. This "reciprocality" has been used in the design of new CSPs by Pirkle *et al.*
(1980) and Pirkle and Finn (1981). For example, the observation that derivatives
of *N*-3,5-DNB-phenylglycine resolve well on fluoro alcohol CSPs **60** and **64** has
been translated into CSPs **65**, **66**, and **67a**. As expected, these CSPs resolve the
enantiomers of 9-anthrylcarbinols. The ionically bonded CSP **67a** is especially
effective. The chemistry employed to make the phenylglycine-derived CSPs has
been also applied to CSPs derived from other amino acids. Ionically bound CSP
67 can be prepared "*in situ*" by passage of a tetrahydrofuran solution of the
amino acid–DNB through an efficient prepacked commercial amino column.

NHDNB   NHDNB   NHDNB
PhĊHCONH(CH₂)₃Si≡   PhĊHCH₂OCONH(CH₂)₃Si≡   RĊHCO₂⁻ N̈H₃(CH₂)₃Si≡

65   **66**   **67 a   R=Ph**

                                                          b   R= i Pr

                                                          c   R= i Bu

As mentioned, the enantiomers of anthrylcarbinols, **22a** and its analogs, were
resolved on CSPs **65–67**. These resolutions and the observed elution orders are
consistent with the chiral recognition model described earlier. The magnitude of
α for these resolutions increases when electron-donating substituents are introduced into the anthryl ring. The ionically bound CSPs **67a–67c** are of great
scope, resolving a large array of arylcarbinols. Table VIII lists chromatographic
data for representative resolutions so obtained. As illustrated by Fig. 8, the
quality of resolution on an analytical CSP (**67a**) is good enough (even at low α
values) to allow accurate assay of enantiomeric purity and absolute configuration
using only a small amount of sample.

In addition to the resolution of arylcarbinols, CSP **67a** (and its analogs) can
separate the enantiomers of solutes within classes **68–85** (Pirkle *et al.*, 1981;
Pirkle and Schreiner, 1981). The interactions used by CSP **67a** for chiral recognition include π–π, hydrogen-bonding (both donor and acceptor), and steric

TABLE VIII
Direct Resolution of Aryl Carbinols on CSP **67a**

$$\underset{Ar}{\overset{HO}{\underset{\diagdown}{\overset{|}{C}}}}\overset{H}{\underset{R}{\diagup}}$$

Ar	R	α^a	Ar	R	α^a
9-Anth	CF$_3$	1.33	2-Naphthyl	CF$_3$	1.08
9-Anth	CH$_3$	1.30	C$_6$H$_5$	CF$_3$	1.06
9-Anth	n-C$_4$H$_9$	1.48	C$_6$H$_5$	CH$_3$	1.05
9-Anth	CH$_2$CO$_2$C$_2$H$_5$	1.27	C$_6$H$_5$	C$_2$H$_5$	1.05
9-Anth	C$_6$H$_5$	1.57	C$_6$H$_5$	CH(CH$_3$)$_2$	1.05
9-(10-CH$_3$)Anth	CF$_3$	1.44	C$_6$H$_5$	C(CH$_3$)$_3$	1.08
9-(10-CH$_3$O)Anth	CF$_3$	1.28	C$_6$H$_5$	n-C$_4$H$_9$	1.08
3-Pyrenyl	CF$_3$	1.08	Mesityl	CH$_3$	1.10
3-Pyrenyl	CH$_3$	1.06	p-Tolyl	CF$_3$	1.08
1-Naphthyl	CF$_3$	1.08	p-Biphenyl	CH$_3$	1.03
1-Naphthyl	CH$_3$	1.14	3,4-Dimethoxyphenyl	CH(CH$_3$)$_2$	1.07

[a]The enantiomer with the absolute configuration shown above is eluted first on (R)-(**67a**).

interactions. Solutes within classes **68–85** have in common the presence of complementary functionality capable of undergoing these interactions. It is no exaggeration to say that a single analytical column of CSP **67a** can be used to determine the enantiomeric purity and absolute configuration of thousands of compounds. Absolute configuration can be determined by the comparison of elution order with that of a known reference compound or can be deduced from a chiral recognition model and the observed elution order. Although it is beyond the scope of this chapter to discuss these chiral recognition models, elution

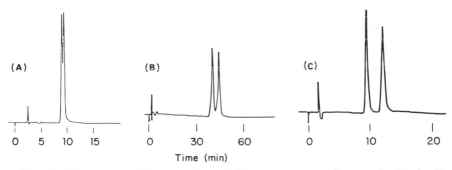

Fig. 8. Chromatographic separation of (A) racemic phenylisopropylcarbinol, (B) racemic trifluoromethyl-α-naphthylcarbinol, and (C) racemic trifluoromethyl-9-anthrylcarbinol on CSP **67a**. From Pirkle and Finn (1981). Copyright 1981 American Chemical Society.

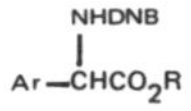

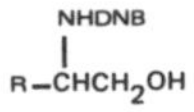

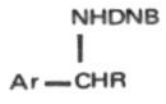

$NHDNB$
$Ar-CHCO_2R$
68

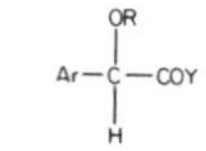

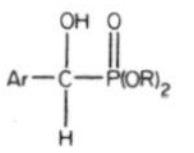

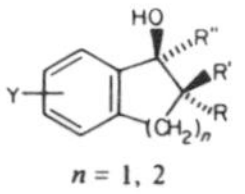

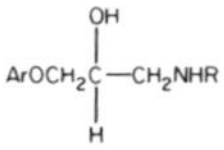

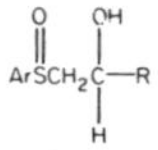

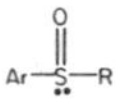

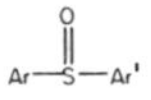

$NHDNB$
$R-CHCH_2OH$
69

$NHDNB$
$Ar-CHR$
70

$Ar-\overset{OR}{\underset{H}{C}}-COY$
71

$Ar-\overset{OH}{\underset{H}{C}}-\overset{O}{P(OR)_2}$
72

$n = 1, 2$
73

$ArOCH_2\overset{OH}{\underset{H}{C}}-CH_2NHR$
74

$ArSCH_2\overset{O}{\overset{\parallel}{C}}-\overset{OH}{\underset{H}{C}}-R$
75

$Ar-\overset{O}{\overset{\parallel}{S}}-R$
76

$Ar-\overset{O}{\overset{\parallel}{S}}-Ar'$
77

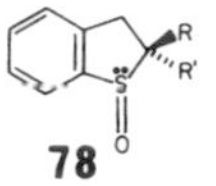

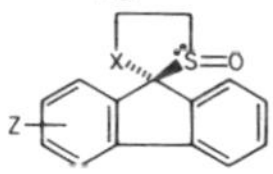

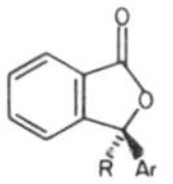

78　　　**79**　　　**80**

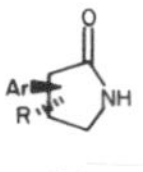

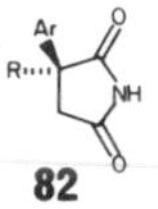

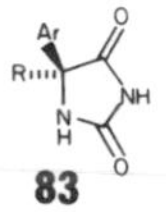

81　　**82**　　**83**

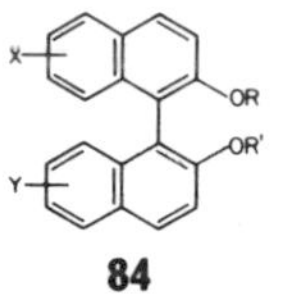

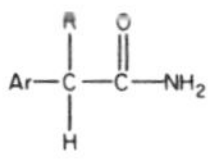

84　　　**85**

orders consistent with one basic three-point model have been observed in every case examined.

Due to the ready availability of (*R*)-phenylglycine and the directness of the synthetic sequence, the preparation of a large quantity of CSP **67a** is easily and inexpensively achieved. A 2 × 30 in. column packed with CSP **67a** prepared on a 40-μm support is capable of resolving a wide variety of racemates on a multigram scale (Finn, 1982). Figure 9 illustrates several gram-sized resolutions obtained on this column. In these cases, only two fractions were collected. Table IX contains data for a representative sampling of the preparative resolutions so obtained. These resolutions have been automated and tens of grams of racemate can be so resolved daily. It should be obvious that a larger quantity of racemate can be resolved in a single run by employing either a larger column or a more efficient column (prepared from smaller silica particles).

iii. 5-ARYLHYDANTOIN CHIRAL STATIONARY PHASE. Among the many solutes that can be resolved on the amino acid-derived CSPs are 5-arylhydantoins. Hydantoin **86**, chosen because of its facile resolvability on the phenylglycine CSP **67a**, has been so resolved preparatively, hydrosilated with triethoxysilane,

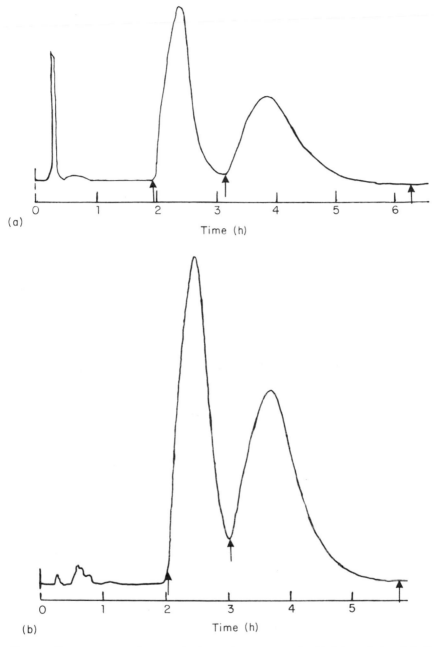

Fig. 9. Preparative resolution of (A) 1.13 g racemic 3−(2′,6′-dimethylnaphthyl)-3-(methyl)phthalide (solute class **80**), (B) 1.60 g racemic 5-(α-naphthyl)-5-(4-pentenyl)hydantoin (solute class **83**), and (C) 1.00 g racemic 2′-methoxy-2-hydroxy-1,1′-binaphthyl (solute class **84**) on a preparative 2 × 30 in. column packed with CSP **67a**.

(*continued*)

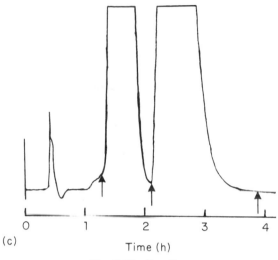

Fig. 9 (*Continued*)

and bonded to silica to afford CSP **87** (W. Pirkle and M. Hyun, unpublished results). A column containing this packing is capable of resolving not only a variety of amino acid–DNB derivatives but also DNB derivatives of amino alcohols and amines.

TABLE IX

Resolution of Racemates on a Preparative Liquid Chromatographic
Column Packed with CSP **67a**

Structure	Amount (g)	α	Enantiomeric excess (%) First band	Second band
	1.13	1.68	>99	>99

TABLE IX (*Continued*)

Structure	Amount (g)	α	Enantiomeric excess (%)	
			First band	Second band

| 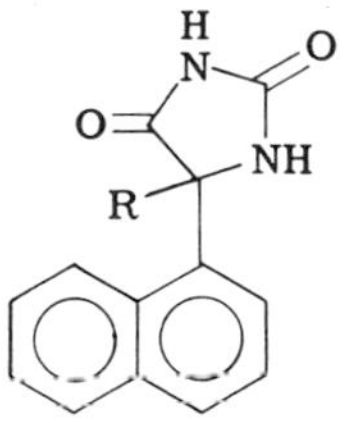 R = $(CH_2)_3CH$=CH_2 | 1.60 | 1.55 | >99 | >99 |
| R = CH_3 | 1.00 | 1.37 | 88 | 92 |

| 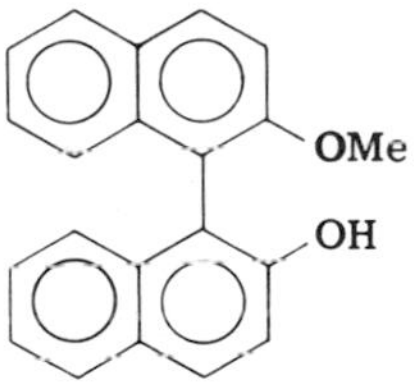 | 1.00 | 1.86 | >99 | >99 |

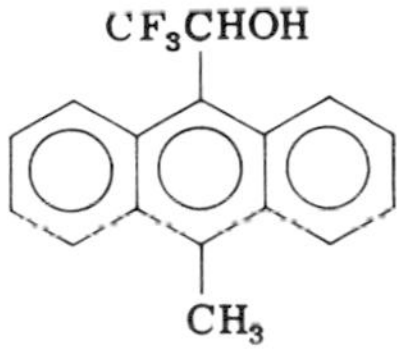	1.00	1.47	>99	>99
	2.00	—	>99	96
	4.00	—	79	85
	8.00	—	73	94

| 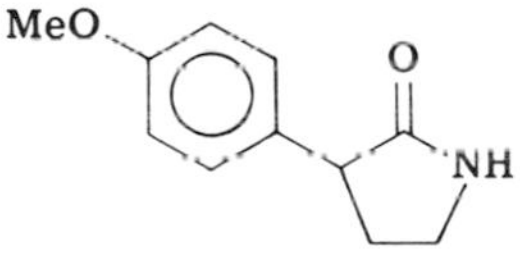 | 1.00 | 1.42 | >99 | 92 |

| 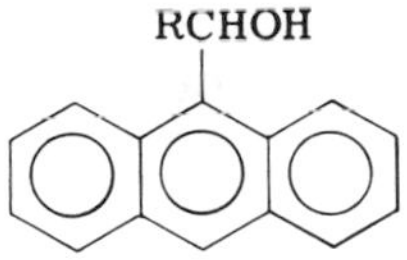 R = CF_3 | 1.00 | 1.46 | 97 | 94 |
| R = n-C_4H_9 | 1.00 | 1.63 | 97 | 99 |

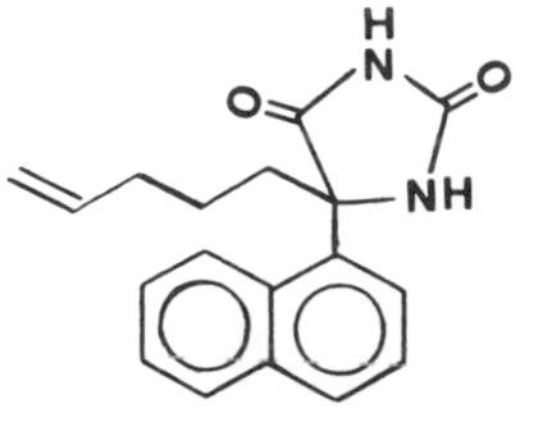

86 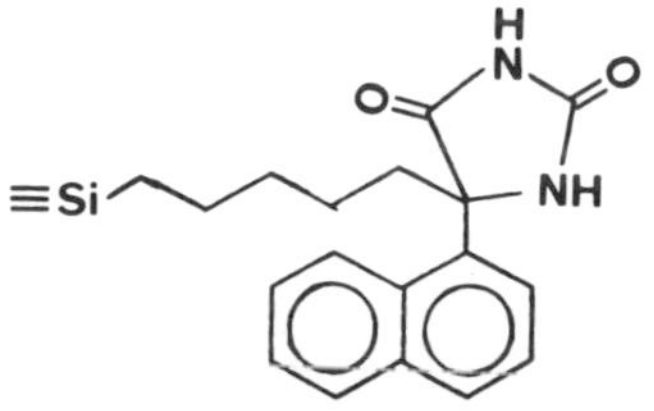   87

iv. DIRECT RESOLUTIONS USING LIGAND-EXCHANGE CHROMATOGRAPHY. The direct resolution of α-amino acids and α-hydroxy acids can be accomplished by ligand-exchange chromatography (LEC). In LEC one has present a metal ion (usually Cu^{2+}, Ni^{2-}, or Zn^{2+}) capable of forming a multidentate complex containing both a chiral ligand and a solute enantiomer. The formation of these diastereomeric complexes is reversible, and solute enantiomers are exchanged rapidly. The chiral ligand can be either bound to a support (CSP) or added to the mobile phase (CMPA). In the former case, enantiomers can be separated when the diastereomeric complexes are of unequal stability. It has been suggested that this stability difference, calculated to be more than 800 kcal/mol in some cases, originates from steric interactions within the multidentate complex. Diastereomeric Cu(II) complexes of bifunctional α-amino acids preferentially adopt the trans-planar structure illustrated by Fig. 10; whenever possible, a second coordination sphere is formed by molecules occupying the axial positions. For steric reasons, only one of the diastereomeric complexes in Fig. 10 can contain an axial solvent molecule. It is suggested that the greater stability of the five-coordinate complex accounts for the observed chiral recognition. This rationale explains the elution order observed for simple bidentate α-amino acids on a proline CSP. With tridentate ligands the elution order is reversed due to the capacity of the S,S-complex to form a five-coordinate complex.

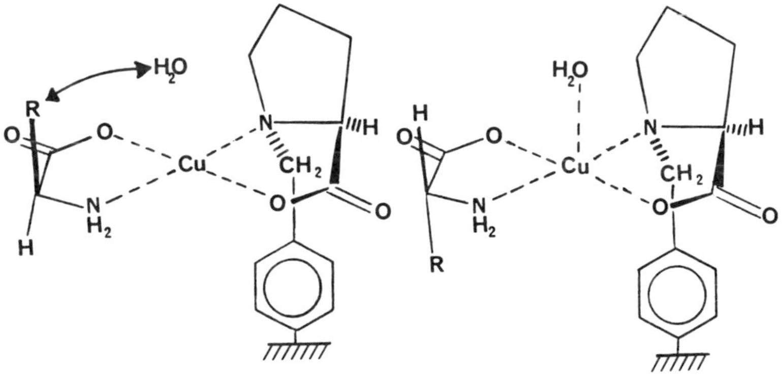

Fig. 10. Typical diastereomeric solvates in an LEC experiment. The R,S-solvate is of higher stability than the S,S-solvate due to the incorporation of an axial solvent ligand.

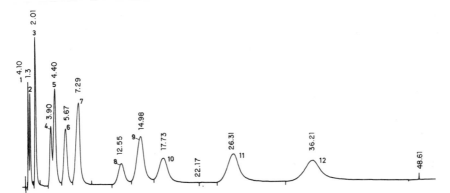

Fig. 11. Separation of six racemic α-amino acids on C_{16}-LHyp-coated LiChrosorb RP-18 5-μm column (100 × 4.2 mm). Eluant: methanol–water (15:85, v/v), 10^{-4} M $CuAc_2$, pH 5.0; flow rate: 2 cm^3/min; column temperature: 293 K; chromatograph: DuPont model 830; detection: UV, 254 nm; elution sequence: (1) LAla, (2) DAla, (3) LNVal, (4) LLeu, (5) LNLeu, (6) DNVal, (7) LPhe, (8) DLeu, (9) LTrp, (10) DNLeu, (11) DPhe, (12) DTrp. The additional numbers in the chromatogram indicate the retention time in minutes. From Davanakov (1980), with permission of the publisher.

Davanakov *et al.* (1979) demonstrated that amino acid-derived CSPs are very effective for the LEC resolution of underivatized α-amino acids. In early studies of Davanakov *et al.* (1979) and of Lefebure *et al.* (1978), the chiral ligands were bound to chloromethylated polystyrene. Although high degrees of stereoselectivity (α > 1.5) were observed for several common α-amino acids, the inefficiency of these columns limited their analytical utility Gübitz *et al.* (1981) circumvented this problem by binding L-proline to modified silica so as to prepare high-efficiency columns. Davanakov *et al.* (1980) developed an alternate approach in which the CSP is prepared "*in situ*" by saturating an efficient, prepacked octadecyl column with hydroxyproline N-alkylated with a fatty alkyl group. The resultant "hydrophobically bound" CSP is successful for the reverse-phase resolution of all common α-amino acids, α values as large as 16.4 being observed. Figure 11 shows a representative chromatogram so obtained. Configurationally consistent elution orders were observed for bidentate α-amino acids. As with all LEC methods, α and κ values are affected by variation in pH, metal ion concentration, and temperature.

c. Chiral Mobile-Phase Additives

i. MECHANISM. When enantiomers are separated on an achiral column by a CMPA, the process of separation is complex. Separation can result from any combination of differential stability of the diastereomeric complexes in the mobile phase, differential adsorption of the diastereomeric complexes themselves, or adsorption of the CMPA on the stationary phase so that the latter behaves as a CSP. The direction of the stereoselectivity afforded by each mechanism need not be the same, and the various mechanisms may compete to degrees that vary with solute structure. Hence, correlation of elution order with absolute configuration may become uncertain if one extrapolates very far from known examples. Indeed, the composition of the mobile phase may play a role in determining elution order.

ii. APPLICATIONS. Pirkle and Sikkenga (1976) partially resolved racemic methyl-2,4-dinitrophenyl sulfoxide on a silica column by the addition of (R)-2,2,2-trifluoro-1-(9-anthryl)ethanol to the CCl_4 mobile phase. The experiment was complicated by detection problems, and only partially resolved samples were obtained. It was subsequently shown that the use of the fluoro alcohol as a CSP rather than as a CMPA afforded superior results for this and similar resolutions.

The majority of the CMPA methods developed involve the separation of α-amino acids by variations on Davanakov's LEC. In all such cases the mobile phase contains a complexing metal ion and a complexable CMPA, usually an amino acid, an amino acid derivative, or a chiral amine. Although column requirements are simple (octadecyl, ion exchange), detection is often a problem and either pre- or postcolumn derivatization is usually necessary. In cases of nonderivatized solutes for which detection of the ternary complex is possible (Gilon et al., 1980), the diastereomeric complexes may have different extinction coefficients. Of the several methods introduced, those of Gil-Av and co-workers (1980) and Lindner et al. (1979) appear to be most useful for the resolution of amino acids.

V. Conclusions

The recent spurt of progress in the chromatographic separation of stereoisomers can leave little doubt as to the direction in which stereochemical analysis and separation is heading rapidly. The next few years will see the commercialization of a variety of chiral chromatography columns, both analytical and preparative. Early indications suggest that there are a large number of potential users for such columns and for improved chiral derivatizing agents and CMPAs.

Acknowledgment

The authors wish to thank the National Science Foundation for support during the writing of this chapter.

References

Ade, E., Helmchen, G., and Heiligenmann, G. (1980). *Tetrahedron Lett.*, p. 1137.

Audebert, R. (1979). *J. Liq. Chromatogr.* **2**, 1063.

Baczuk, R., Landram, G., Dubois, R., and Dehm, H. (1971). *J. Chromatogr.* **60**, 351.

Bergot, B., Anderson, R., Schooley, D., and Henrick, C. (1978). *J. Chromatogr.* **155**, 97.

Blaschke, G. (1980). *Angew. Chem. Int. Ed. Engl.* **19**, 13.

Blaschke, G., and Markgraf, H. (1980). *Chem. Ber.* **113**, 2031.

Blaschke, G., Kraft, H., and Markgraf, H. (1980). *Chem. Ber.* **113**, 2318.

Boding, H., and Musso, H. (1978). *Angew. Chem. Int. Ed. Engl.* **17**, 851.

Boyd, P., Gadaginamath, G., Hamilton, R., Yagi, H., and Jerina, D. (1978). *Tetrahedron Lett.*, p. 2487.

Buss, P., and Vermeulen, T. (1968). *Ind. Eng. Chem.* **60**, 12.

Corey, E., Danheiser, R., Chandrasekaren, S., Keck, G., Gopalan, B., Larsen, S., Siret, P., and Gras, J. (1978). *J. Am. Chem. Soc.* **100**, 8034.

Corey, E. J., Hopkins, P., Kim, S., Yoo, S., Nambiar, K., and Falck, J. (1979). *J. Am. Chem. Soc.* **101**, 7131.

Dalgleish, C. (1952). *J. Chem. Soc.* **137**, 3940.

Davanakov, V. (1980). *Adv. Chromatogr.* **18**, 139.

Davanakov, V., Zolotarev, Y., and Kurganov, A. (1979). *J. Liq. Chromatogr.* **2**, 1191.

Davanakov, V., Bochkov, A., Kurganov, A., Roumeliotis, P., and Unger, K. (1980). *Chromatographia* **13**, 677.

Dotsevi, G., Sogah, Y., and Cram, D. (1979). *J. Am. Chem. Soc.* **101**, 3035.

Eberhardt, R., Glotzmann, C., Lehner, H., and Schlögl, K. (1974). *Tetrahedron Lett.*, p. 4365.

Enders, P., and Lotter, H. (1981). *Angew Chem. Int. Ed. Engl.* **20**, 795.

Farnum, D., Yeysoglu, T., Cardé, A., Duhl-Emswiler, B., Pancoast, T., Reitz, T, and Cardé, R. (1977). *Tetrahedron Lett.*, p. 4009.

Finn, J. (1982). Ph.D. Thesis, Univ. of Illinois, Urbana.

Gil-Av, E., Tishbee, A., and Hare, P (1980) *J Am Chem Soc* **102**, 5115

Gilon, C., Leshem, R., and Grushka, E. (1980). *Anal. Chem.* **52**, 1206.

Goto, J., Hasegawa, M., Sctsoko, H., Himada, K., and Nambara, T. (1978). *J. Chromatogr.* **152**, 413.

Gübitz, G., Vellenz, W., and Santi, W. (1981). *J. Liq. Chromatogr.* **4**, 701.

Guyon, F., Oliveros, L., and Caude, M. (1978). *J. Chromatogr.* **152**, 551.

Häkli, H., Mintas, M., and Mannschreck, A. (1979). *Chem. Ber.* **112**, 2028.

Hara, S., and Dobashi, A. (1979a). *J. Liq. Chromatogr.* **2**, 883

Hara, S., and Dobashi, A. (1979b). *J. Chromatogr.* **186**, 543.

Helmchen, G., and Nill, G. (1979). *Angew. Chem Int Ed Engl* **18**, 65

Helmchen, G., and Strubert, W. (1974). *Chromatographia* **7**, 713.

Helmchen, G., Ott, R., and Sauber, K. (1972). *Tetrahedron Lett.*, p. 3873.

Helmchen, G., Völter, H., and Schühle, W. (1977). *Tetrahedron Lett.*, p. 1417.

Helmchen, G., Nill, G., Flockerzi, D., Schühle, W., and Youssef, S. (1979a). *Angew Chem. Int. Ed. Engl.* **18**, 62.

Helmchen, G., Nill, G., Flockerzi, D., and Youssef, S. (1979b). *Angew. Chem. Int. Ed. Engl.* **18**, 63.

Hesse, G., and Hagel, R. (1976). *Liebig's Ann. Chem.*, p. 996.

Imai, K., and Marumo, S. (1976). *Tetrahedron Lett.*, p. 1211.

Koreeda, M., Weiss, G., and Nakanishi, K. (1973). *J. Am. Chem. Soc.* **95**, 239.

Krull, I. (1977). *Adv. Chromatogr.* **16**, 176.

Kuritani, H., Iwata, F., Sumiyoshi, M., and Shingu, K. (1977). *J. Chem. Soc. Chem. Commun.*, p. 542.

Lefebure, B., Audebert, R., and Quivoron, C. (1978). *J. Liq. Chromatogr.* **1**, 761.

Leitch, R., Rothbart, H., and Rieman, W. (1968). *Talanta* **15**, 213.

Lindner, W., LePage, J., Davies, G., Seitz, P., and Karger, B. (1979). *J. Chromatogr.* **185**, 323.

Lochmüller, C., and Ryall, R. (1978). *J. Chromatogr.* **150**, 511.

Lochmüller, C., and Souter, R. (1975). *J. Chromatogr.* **113**, 283.

McKay, S., Mallen, D., Schobsall, P., Swann, B., and Williamson, W. (1979). *J. Chromatogr.* **170**, 482.

Meyers, A., and Slade, J. (1980). *J. Org. Chem.* **45**, 2912.

Meyers, A., Slade, J., Smith, R., Mihelich, E., Hershenson, F., and Liang, C. (1979). *J. Org. Chem.* **44**, 2247.

Mikes, F., and Boshart, G. (1978). *J. Chem. Soc. Chem. Commun.*, p. 173.

Mikes, F., Boshart, G., and Gil-Av, E. (1976). *J. Chromatogr.* **122**, 205.

Mintas, M., Mannschreck, A., and Klasinc, L. (1981). *Tetrahedron* **37**, 867.

Mori, K., and Sasaki, M. (1980). *Tetrahedron* **36**, 2197.

Nambara, T., Megawa, S., Hasegawa, M., and Goto, J. (1978). *Anal. Chim. Acta* **101**, 111.

Pappo, R., Collins, P., and Jung, C. (1973). *Tetrahedron Lett.*, p. 943.

Pirkle, W., and Adams, P. (1979). *J. Org. Chem.* **44**, 2169.

Pirkle, W., and Adams, P. (1980a). *J. Org. Chem.* **45**, 4111.

Pirkle, W., and Adams, P. (1980b). *J. Org. Chem.* **45**, 4117.

Pirkle, W., and Boeder, C. (1978a). *J. Org. Chem.* **43**, 1950.

Pirkle, W., and Boeder, C. (1978b). *J. Org. Chem.* **43**, 2091.

Pirkle, W., and Finn, J. (1981). *J. Org. Chem.* **46**, 2935.

Pirkle, W., and Hauske, J. (1977a). *J. Org. Chem.* **42**, 1839.

Pirkle, W., and Hauske, J. (1977b). *J. Org. Chem.* **42**, 2781.

Pirkle, W., and Hoekstra, M. (1974). *J. Org. Chem.* **39**, 3904.

Pirkle, W., and House, D. (1979). *J. Org. Chem.* **44**, 1957.

Pirkle, W., and Rinaldi, P. (1978). *J. Org. Chem.* **43**, 3803.

Pirkle, W., and Rinaldi, P. (1979). *J. Org. Chem.* **44**, 1025.

Pirkle, W., and Schreiner, J. (1981). *J. Org. Chem.* **46**, 4988.

Pirkle, W., and Sikkenga, D. (1975). *J. Org. Chem.* **40**, 3430.

Pirkle, W., and Sikkenga, D. (1976). *J. Chromatogr.* **123**, 400.

Pirkle, W., and Simmons, K. (1982). *J. Org. Chem.* **48**.

Pirkle, W., House, D., and Finn, J. (1980). *J. Chromatogr.* **192**, 143.

Pirkle, W., Finn, J., Schreiner, J., and Hamper, B. (1981). *J. Am. Chem. Soc.* **103**, 3964.

Robertson, M. (1981). Ph.D. Thesis, Univ. of Illinois, Urbana.

Schmid, G., Kokuyama, T., Aksaka, K., and Kishi, Y. (1979). *J. Am. Chem. Soc.* **101**, 259.

Schwanghart, A., Backmann, W., and Blaschke, G. (1977). *Chem. Ber.* **110**, 778.

Scott, C., Petrin, M., and McCorkle, T. (1976). *J. Chromatogr.* **125**, 157.

Sousa, L., Sogah, G., Hoffman, D., and Cram, D. (1978). *J. Am. Chem. Soc.* **100**, 4569.

Souter, R. W. (1976). *Chromatographia* **9**, 635.

Stewart, K., and Doherty, R. (1973). *Proc. Natl. Acad. Sci. USA* **70**, 2850.

Szokan, G., Ruff, F., and Kocsmann, A. (1980). *J. Chromatogr.* **198**, 207.

Tamegai, T., Ohmae, M., Kanabe, K., and Tomoeda, M. (1979). *J. Liq. Chromatogr.* **2**, 1229.

Valentine, D., Chan, K., Scott, C., Johnson, K., Toth, K., and Saucy, G. (1976). *J. Org. Chem.* **41**, 62.

Westly, J., Halpern, B., and Karger, B. (1968). *Anal. Chem.* **40**, 2046.

Williams, D., and Phillips, J. (1981). *J. Org. Chem.* **46**, 5452.

Wulff, G., Vesper, W., Grobe-Einsler, R., and Sarhan, A. (1977). *Makromol. Chem.* **178**, 2817.

Yuki, H., Okamoto, Y., and Okamoto, I. (1980). *J. Am. Chem. Soc.* **102**, 6356.

7

Nuclear Magnetic Resonance Analysis Using Chiral Derivatives

Shozo Yamaguchi
Department of Chemistry
College of General Education
Tohoku University
Sendai, Japan

I. Introduction

The determination of enantiomeric composition by the measurement of optical rotation appears to be very simple, but in practice it must be done with great care if a correct value is to be obtained (see Chapter 2, this volume). There are several major potential sources of error:

ISBN 0-12-507701-7

1. If the maximum rotation of the compound is unknown, it must be determined by some means, most often in the past by a tedious resolution. Even
   then, the value may be in doubt because of the possibility of incomplete
   resolution. If the maximum rotation has been published, one must rely on
   the accuracy of the published value. Scrupulous care must be exercised
   during the measurement to make certain that the observed rotation and
   maximum rotation are determined under identical conditions of solvent,
   concentration (Reich, 1976), temperature, and pH (Horn and Pretorius,
   1954; Klyne and Buckingham, 1978), because all of these may have a
   strong influence on the value. Sometimes there are large deviations from
   linearity between optical rotation and these variables (Horeau, 1969); a
   determination that assumes such linearity may be in error.
2. The accuracy of the determination is also dependent on the magnitude of
   the optical rotation, which may be very low in some cases, such as for
   compounds that are chiral by virtue of deuterium substitution.
3. There is always the problem of the presence of a chiral impurity, which
   will cause an unsuspected error in the observed rotation. Therefore, simple
   and reliable alternate methods of determining enantiomeric purity continue
   to be explored.

Two associated parameters are necessary for the determination of enantiomeric composition (Raban and Mislow, 1967). One is a structurally dependent
property due to the differences between members of an enantiomeric pair or to
differences between diastereomeric derivatives prepared quantitatively from the
mixture of enantiomers. The other is a parameter associated with the first, which
can be quantitatively related to the amount of each individual stereoisomer.

II. Application of Chemical Shift
Nonequivalence

The most widely used methods of determining enantiomeric ratio are chromatographic (Schurig *et al.*, 1978, Helmchen *et al.*, 1979a,b); and NMR methods (Raban and Mislow, 1967, Gaudemer, 1977). In the NMR method the
parameters are the nonequivalent chemical shifts of selected signals from diastereotopic groups and the relative intensities of these signals. In order to
discriminate between enantiotopic groups (Jacobus *et al.*, 1968) in enantiomeric
mixtures they must be rendered externally diastereotopic by the use of chiral
lanthanide shift reagents (McCreary *et al.*, 1974) or chiral solvents (Pirkle and
Sikkenga, 1977). Alternatively an enantiomeric mixture is converted to a pair of
diastereomers with an appropriate chiral derivatizing reagent, and then the enan-

tiotopic groups in the original sample are observed as diastereotopic groups by internal comparison (Raban and Mislow, 1967; Gaudemer, 1977).

A. Proton Nuclear Magnetic Resonance Chemical Shift Nonequivalence

O-Methylmandelic acid (**1**) has been used (Jacobus *et al.*, 1968; Sandman *et al.*, 1968) as a chiral derivatizing agent for determining the enantiomeric ratio of secondary carbinols as well as primary carbinols that are chiral by virtue of deuterium substitution. Diastereomeric *O*-methyl mandelates of (*R*)- and (*S*)-methylphenylcarbinols [e.g., the diastereomeric mixture of (*R,R*)- and (*R,S*)-esters] have diastereotopic methyl groups that give NMR signals with different chemical shifts (Raban and Mislow, 1967, Dale and Mosher, 1968). In this case the two parameters are the different chemical shifts and the relative intensities of the signals, which will be directly proportional to the ratio of *R,R*- and *R,S*-diastereomers present in the mixture. In order that the ratio of diastereomeric derivatives correctly measures the enantiomeric composition of the substrate [in this case (*R*)- and (*S*)-methylphenylcarbinols] the following conditions must be satisfied. (*a*) The reagent must be stable to racemization under the conditions of the derivatization process. (*b*) The reaction to form diastereomeric derivatives must be quantitative with respect to the enantiomeric substrates in order to avoid spurious results due to kinetic resolution. (*c*) There must be no concentration of one diastereomer over the other in any purification step such as fractional crystallization of the diastereomer mixture. (*d*) The reagent must carry an appropriate probe (or probes) that affords clearly separated NMR signals such as those from a methyl, methoxyl, or *tert*-butyl group for ^{1}H NMR or a CF_3 group for ^{19}F NMR. The successful derivatizing agent for the development of distinct diastereotopic chemical shift nonequivalences should have a well-defined conformational preference so that the diastereotopic groups will be located in distinctly different magnetically anisotropic environments. The presence of a benzene ring and functional groups that exert an effective shielding or deshielding influence on the diastereotopic probe groups is clearly advantageous.

1. THE MOSHER REAGENT, α-METHOXY-α-TRIFLUOROMETHYLPHENYLACETIC ACID, AS A CHIRAL DERIVATIZING REAGENT

Mosher and co-workers (Dale and Mosher, 1968) examined the application and limitations of a series of α-substituted phenylacetic acids (1–5) as chiral derivatizing reagents for NMR analysis and found that these reagents are prone to racemization, especially on reaction with hindered carbinols by virtue of their α-hydrogens. On the basis of these observations they developed (Dale *et al.*, 1969; Dale and Mosher, 1973) an extremely useful chiral derivatizing reagent: α-methoxy-α-trifluoromethylphenylacetic acid [MTPA (6), the Mosher reagent]. This reagent has several advantages. (*a*) It is remarkably stable toward racemization even under severe conditions of acidity, basicity, and temperature because it does not carry an α-hydrogen. (*b*) Derivatives of MTPA generally show substantial chemical shift differences in the signals of the diastereotopic groups. (*c*) The presence of the CF_3 group also allows the use of ^{19}F NMR. (*d*) Because MTPA derivatives are quite volatile and soluble, they can be analyzed by gas–liquid chromatography (GLC) and high-pressure liquid chromatography (HPLC) as well as by NMR methods. A wide variety of carbinols and amines react smoothly with MTPA chloride[1] to give the corresponding esters and amides. Highly hindered carbinols may require special conditions to force the reaction to completion. In practice, the chiral derivatizing reagent is used in excess in order to force the reaction to completion. 4-Dimethylaminopyridine has been recommended (Kabuto *et al.*, 1980; H. S. Mosher, personal communication) as an acylation catalyst.

Ph—CHCOOH
|
X

1 X = OMe

2 X = *t* Bu

3 X = CF₃

4 X = OH

5 X = Cl

Ph
|
MeO
C—COOH
F₃C
6

Figure 1 shows the 60-MHz NMR spectra of the methyl-*tert*-butylcarbinyl esters of (R)-MTPA (7) and (S)-O-methylmandelic acid (8) formed from methyl-*tert*-butylcarbinol that is enriched in the R-isomer to the extent of 7.8 and

[1]The optically active chloride can easily be prepared (91% yield) from the potassium salt of (R)-MTPA and oxalyl chloride in the presence of 18-crown-6 in benzene solution at room temperature (H. S. Mosher, personal communication).

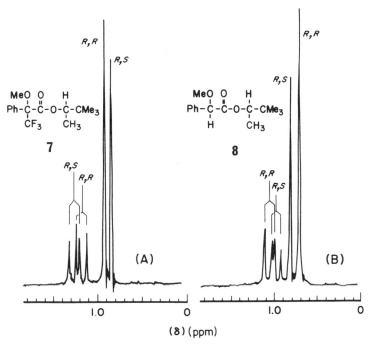

Fig. 1. (A) 60-MHz NMR spectra of (R)-MTPA esters from 7.8% ee methyl-*tert*-butylcarbinol. (B) 60-MHz NMR spectra of (S)-O-methyl mandelate esters from 11.7% ee methyl-*tert*-butylcarbinol. From Dale and Mosher (1973). Copyright 1973 American Chemical Society

11.7%, respectively. Spectrum (A) in Fig. 1 clearly indicates that the *tert*-butyl signal from the R,R-diastereomer of the MTPA esters appears at lower field than that of the R,S-diastereomer. However, the methyl signal from the R,R-diastereomer resonates at higher field than that of the R,S-isomer. These chemical shift nonequivalences of the *tert*-butyl and methyl groups are produced by a selective shielding due to the α-phenyl group in the MTPA moiety. The relative peak areas of the well-separated *tert*-butyl signals are a measure of the diastereomeric composition and thus of the enantiomeric purity of the original carbinol. The accuracy of the percent enantiomeric excess (% ee) determined in this manner will depend on the NMR instrument, the tuning of the instrument, and the signal separation. Under fair conditions the value should be within ±1% but can easily be much better or worse. The maximum specific rotation of methyltrifluoromethylcarbinol has been determined on the basis of the observed specific rotation of a sample of the carbinol, and its enantiomeric purity has been determined by the present method (Feigl, 1965). This method is also applicable

TABLE I
^{1}H-NMR Chemical Shift Differences for MTPA Derivatives
9A and 9B[a]

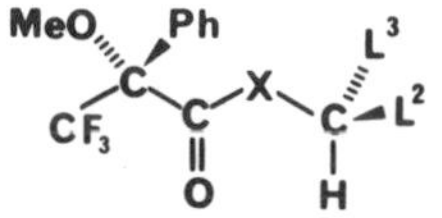

X = O (NH)

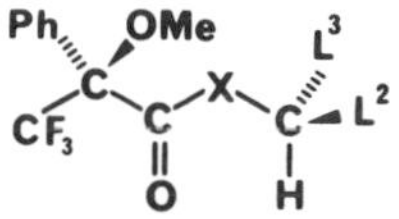

Ligand		Chemical shift difference Δδ (ppm)	
L^3	L^2	L^3 (9A – 9B)	L^2 (9A – 9B)
Et	Me	+0.10 (+0.07)[b]	−0.13 (−0.07)[b]
iPr	Me	+0.08	−0.08
tBu	Me	+0.07	−0.07
tBu	nBu	+0.06	—
Ph	Me	—	−0.06 (−0.07)[b]
Ph	tBu	—	−0.05
COOMe	Me	—	(−0.08)[b]

[a]Data from Dale and Mosher (1973).
[b]Values in parentheses are those for the MTPA amides.

to tertiary carbinols (Dale and Mosher, 1973; Mukaiyama *et al.*, 1978) when a quantitative acylation of the carbinols can be performed.

A useful NMR configurational correlation scheme (Dale and Mosher, 1973) for assigning the absolute configuration of carbinols and amines has been developed from a detailed investigation of MTPA derivatives. As shown in Table I, the ^{1}H-NMR signal due to group L^3 in structure 9A appears consistently at lower field than that of L^3 in structure 9B [ΔδL^3 (9A − 9B) > 0], whereas the signal due to group L^2 in structure 9A appears at higher field than that of L^2 in structure 9B [ΔδL^2 (9A − 9B) < 0]. If according to the Cahn–Ingold–Prelog nomenclature scheme the group priority order at the carbinyl carbon is O > L^3 > L^2 > H, then structures 9A and 9B possess R,R- and S,R-configurations, respectively. This correlation scheme allows the configurational assignment of (R)-MTPA derivatives (R,R- and R,S-pairs) prepared from partially active carbinols and

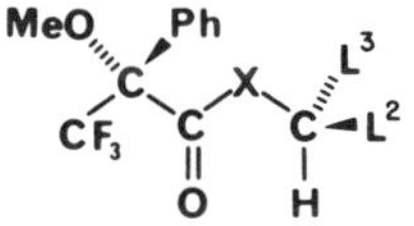

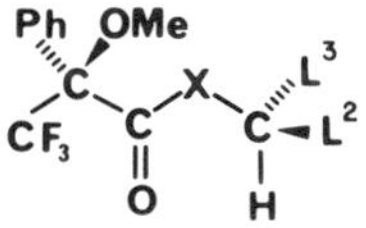

X = O or NH

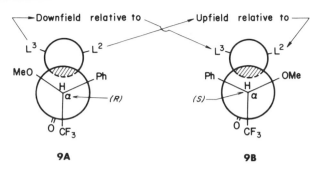

Fig. 2. Conformational correlation model for (R)-MTPA and (S)-MTPA derivatives **9A** and **9B**, respectively.

amines to be accomplished along with simultaneous determination of the % ee of the original carbinols (or amines). This useful NMR configurational correlation scheme can be reasonably interpreted by the model depicted in Fig. 2. The following observations support the proposed model.

It has been concluded from chiroptic studies (Barth *et al.*, 1970) that a preferred conformation of MTPA esters has the α-CF$_3$ and carbonyl groups of the MTPA moiety in an eclipsed arrangement. The conformational model represented by **9A** and **9B** has this feature, and it also has the carbinyl hydrogen of the carbinol moiety eclipsed with the ester carbonyl on the other side. The preferred conformation of the ester group has been discussed before (Mathieson, 1965) and corroborated by several investigations (Helmchen, 1974; Helmchen *et al.*, 1977; Pirkle and Hauske, 1976). Although the differences in chemical shifts for the internally diastereotopic α-OMe groups in MTPA esters and amides are generally small (0–0.08 ppm) (Dale and Mosher, 1973), those for MTPA derivatives of α-amino esters (**10**) [Δδ(OMe) (R,R − R,S) = 0.17–0.23 ppm] (Yasuhara *et al.*, 1978) and α-hydroxy esters (**11**) (Yasuhara and Yamaguchi, 1980) are considerably larger and can be used for making configurational assignments. If one assumes that the crucial conformation for the R,R-diastereomer is that shown in structure **10** or **11**, in which the COOMe group is preferentially oriented toward the OMe group, and also assumes that the COOMe group exerts a deshielding effect (Yasuhara and Yamaguchi, 1980) on the facing OMe group, then the above stereochemical correlation scheme can be rationalized by the Mosher model (Dale and Mosher, 1973) represented by **9A** and **9B**. This NMR

MeO,,,,   Ph
     C   X,,,,COOMe
CF$_3$ C   C←R
         ‖ |
         O   H

10 X = NH   **11** X = O

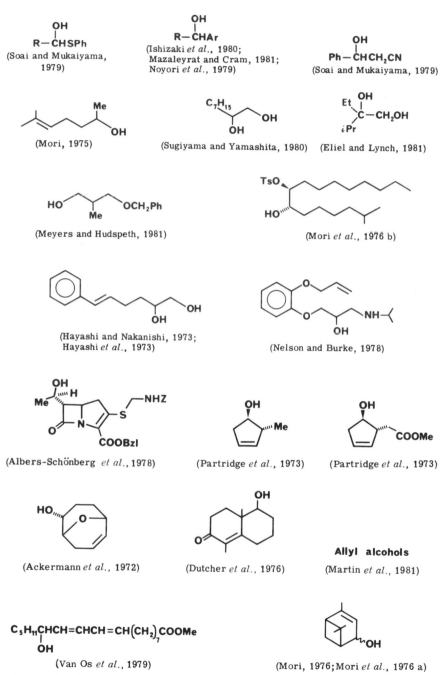

Scheme 1. Some compounds for which configurations or enantiomeric compositions have been determined using MTPA and ^{1}H-NMR spectroscopy.

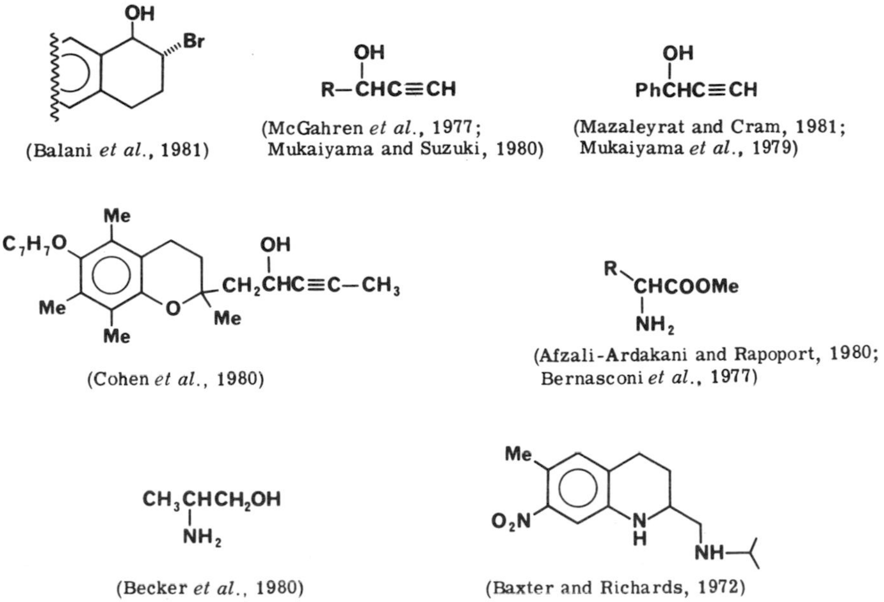

(Balani *et al.*, 1981)

$$R-\overset{OH}{\underset{|}{CH}}C\equiv CH$$

(McGahren *et al.*, 1977;
Mukaiyama and Suzuki, 1980)

$$Ph\overset{OH}{\underset{|}{CH}}C\equiv CH$$

(Mazaleyrat and Cram, 1981;
Mukaiyama *et al.*, 1979)

(Cohen *et al.*, 1980)

$$\overset{R}{\underset{|}{\underset{NH_2}{\diagdown}}}CHCOOMe$$

(Afzali-Ardakani and Rapoport, 1980;
Bernasconi *et al.*, 1977)

$$\overset{CH_3\overset{|}{\underset{|}{CH}}CH_2OH}{\underset{NH_2}{}}$$

(Becker *et al.*, 1980)

(Baxter and Richards, 1972)

Scheme 1 (*Continued*)

method, based on derivatization by MTPA, has been used extensively for the determination of the enantiomeric purity and absolute configuration of various carbinols, amines, and amino alcohols. A number of additional examples are shown in Scheme 1.

The sensitivity of NMR analysis has improved dramatically with the wider use of Fourier transform and 200- to 500-MHz NMR. In general, the required amount of MTPA derivative can be reduced to 0.5 mg or less, and satisfactory spectra can still be obtained (see Fig. 4, Section III,A).

An NMR configurational correlation model for diastereomeric O-methyl mandelate esters (12A and 12B) that explains the observed chemical shift non-equivalence (e.g., for the methyl *tert*-butyl ester, Fig. 1B) has been proposed (Fig. 3) (Dale and Mosher, 1973). The use of this model for determining the absolute configurations of 13 (Trost and Curran, 1981) and 14 (Trost *et al.*, 1980) has been reported.

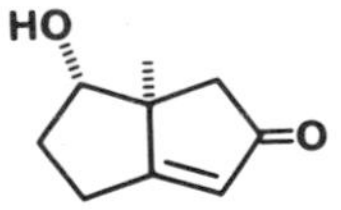

13

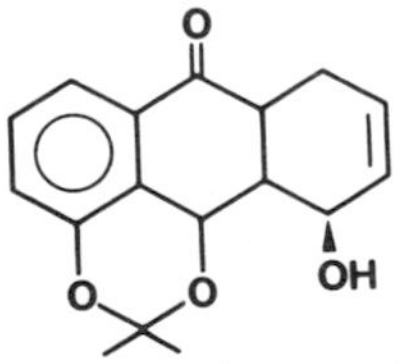

14

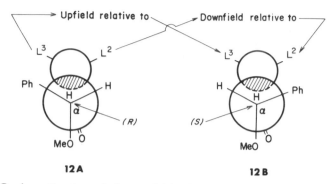

Fig. 3. Conformational correlation model for *O*-methyl mandelate esters **12A** and **12B** (Dale and Mosher, 1973).

2. CHIRAL REAGENTS OTHER THAN MTPA

In addition to the various α-substituted phenylacetic acids described above, chiral α-phenylbutyric acid has also proved to be a useful derivatizing agent for assigning the absolute configuration of secondary carbinols, amines (Helmchen, 1974; Helmchen *et al.*, 1972, 1977), and mercaptans (Helmchen and Schmierer, 1976) using ^{1}H-NMR spectroscopy. In the empirical NMR configurational correlation model proposed, the diastereomer with the signal for R_1 that resonates at relatively higher field corresponds to configuration **15A**, whereas the other diastereomer, the R_1 signal of which appears at lower field, corresponds to configuration **15B**.

$$X = O, NH, S$$

The enantiomeric purity and absolute configuration of chiral α-deuterated benzylamine has been determined by the conversion of this compound to the diastereomeric 2-thioxo-1,3-oxazolidine derivatives (Gerlach, 1966; Gaudemer, 1977). Also, *l*-methoxyacetyl chloride (**16**) (Galpin and Huitric, 1968; Cochran and Huitric, 1971) and *d*-camphor-10-sulfonyl chloride (**17**) (Hoyer *et al.*, 1972) have been reported to be useful chiral derivatizing reagents for determining enantiomer ratios and absolute configurations of carbinols and amines. The methylene protons of the —CH$_2$—CO— group in **16** and those in the methyl and methylene groups in **17** are useful probes.

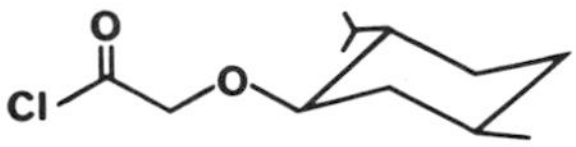

16 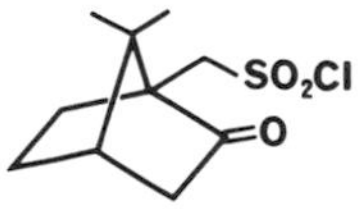   **17**

Although there are several examples of determining the enantiomeric purity of chiral carbonyl compounds by NMR using chiral lanthanide shift reagents (Fraser *et al.*, 1971; Goering *et al.*, 1974; McCreary *et al.*, 1974; Felsky *et al.*, 1978), stable diastereomeric derivatives have been studied less extensively for this purpose. It has been reported (Meyers and Birch, 1979) that the C-1 proton of the aldimine (**20**), derived from partially active α-substituted aldehydes (**18**) and enantiomerically pure 2-amino-1-methoxymenth-8-ene (**19**), exhibits chemical shift nonequivalence ($\Delta\delta \approx 0.17$ ppm) in its ^{1}H-NMR spectrum. This difference is sufficient to permit the determination of the enantiomer ratio of such aldehydes. The presence of excess reagent **19** does not interfere with the NMR analysis.

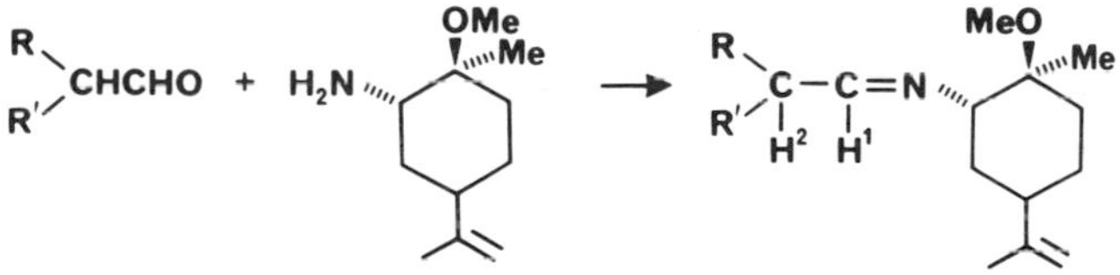

18   **19**   **20**

The chiral reagent α-methyl-α-methoxypentafluorophenylacetic acid [MMPA (**21**)] was prepared and studied as a chiral derivatizing reagent for various amines. It is possible to analyze the resulting diastereomeric amides by both ^{1}H NMR (Pohl and Trager, 1973) and ^{19}F NMR (Nicholas *et al.*, 1973). The X-ray crystallographic analysis (Valente *et al.*, 1980) of the MMPA amide (**22**) of (*S*)-phenylethylamine shows that the conformation in the crystalline state has the carbinyl hydrogen and the amide carbonyl group approximately eclipsed with a hydrogen bond between the NH and OMe groups. On the basis of this solid-state conformation and the observed chemical shift non-equivalence for diastereomeric amides, an NMR configuration model closely similar to that of MTPA derivatives has been proposed.

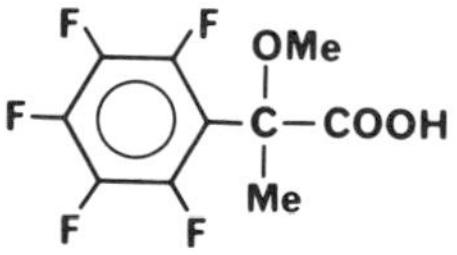

21 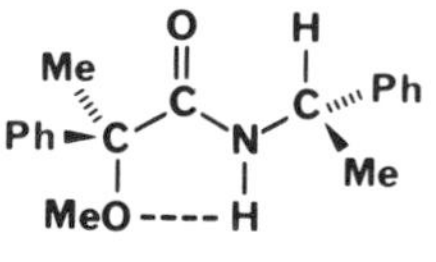   **22**

(*R*)-[1-(9-Anthryl-2,2,2-trifluoroethoxy]acetic acid (**23**) is reported to be a versatile chiral reagent for determining the absolute configuration and enantiomeric purity of alcohols, thiols, and amines using ^{1}H-NMR analysis (Pirkle and Simmons, 1981).

X = O, NH, S

A conformational model has been proposed for assigning the absolute configuration. It is based on the sense of the chemical shift nonequivalence for R_1 and R_2 groups in the diastereomeric derivatives **24**. The NMR chemical shift differences between diastereomeric carbamates (Pirkle and Simmons, 1979) also can be correlated with the stereochemistry at the carbinyl center of the original carbinol. Optically active perfluoro-2-propoxypropionic acid was prepared, and its use in ^{1}H- and ^{19}F-NMR spectroscopy for determining the enantiomeric purity of alcohols has been examined (Kawana and Ishikawa, 1980).

B. *Fluorine Nuclear Magnetic Resonance Chemical Shift Nonequivalence*

^{19}F-NMR chemical shift nonequivalence has been observed for the α-CF$_3$ group of diastereomeric MTPA derivatives (Dale and Mosher, 1968; Dale *et al.*, 1969). The ^{19}F-NMR signal has several advantages over the ^{1}H-NMR signal. (*a*) The signal appears with simple form and high intensity in an uncongested region uncomplicated by overlapping signals, and (*b*) the ^{19}F chemical shift differences for the α-CF$_3$ group of such diastereomeric derivatives are generally much greater (0.11–0.71 ppm) than the chemical shift differences of the proton signals (0.03–0.13 ppm) in the same compounds. Therefore, determinations of enantiomeric purity based on the ^{19}F resonances are, in many cases, more reliable than those based on ^{1}H-NMR spectroscopy. Detailed investigations with various substrates clearly indicate that the ^{19}F signal due to the MTPA derivative with configuration **9A** generally appears at higher field than that with configuration **9B** (Table II) (Sullivan *et al.*, 1973). Several representative examples of ^{19}F chemical shift nonequivalence for the MTPA diastereomers are summarized in Table II. A configurational correlation model for rationalizing the observed chemical shift non-equivalence of ^{19}F signals of these diastereomeric MTPA derivatives has been proposed (Sullivan *et al.*, 1973). This MTPA–^{19}F-NMR

TABLE II
^{19}F-NMR Chemical Shift Differences
for MTPA Derivatives 9A and 9B[a]

9A

9B

$X = O(NH)$

Ligand		Chemical shift difference $\Delta\delta$ (ppm),
L^3	L^2	CF$_3$ (9A − 9B)
iPr	Me	+0.18
tBu	Mc	+0.22
COOEt	Me	+0.45
Ph	Me	+0.20 (+0.25)[b]
Ph	tBu	+0.43
l-Menthyl		+0.12 (+0.10)[b]
Et	Me	0 (+0.10)[b]

[a]Data from Sullivan et al. (1973).
[b]Values in parentheses are those for MTPA amides.

method has been applied to many diverse types of compound; for example, it has been used for determining the epimeric ratio of dihydroneoergosterol (25) (Felsky

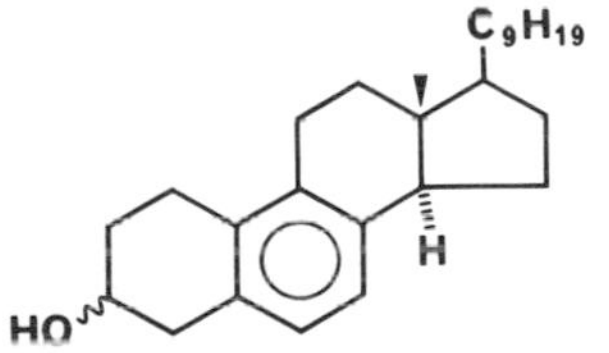

25

et al., 1978). Another interesting application (Lee and Keana, 1978) is the identification of nitroxide diastereomers. The spectrum for the (R)-MTPA derivative of 26A showed two signals ($\Delta\delta CF_3 = 8.5$ Hz), whereas 26B showed only one, indicating that 26A and 26B possess trans- and cis-configurations, respectively. Other compounds whose configurations or enantiomeric compositions have been determined by the ^{19}F-NMR method are illustrated in Scheme 2. α-Methyl-α-methoxypentafluorophenylacetic acid (Nicholas et al., 1973) and perfluoro-2-propoxypropionic acid (Kawana and Ishikawa, 1980) also have been used as chiral derivatizing agents for ^{19}F-NMR analysis.

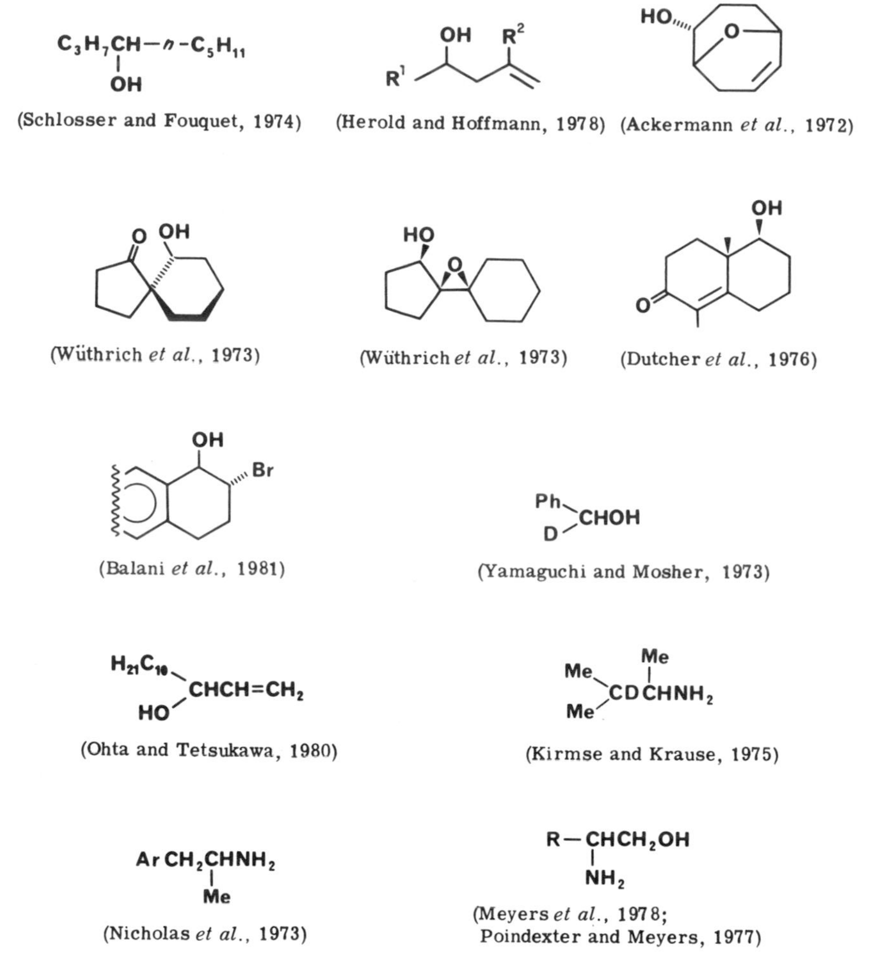
C3H7CH—n-C5H11
OH
(Schlosser and Fouquet, 1974)
OH R2
R1
(Herold and Hoffmann, 1978)
HO
O
(Ackermann et al., 1972)
O OH
(Wüthrich et al., 1973)
HO
O
(Wüthrich et al., 1973)
OH
O
(Dutcher et al., 1976)
OH
Br
(Balani et al., 1981)
Ph
CHOH
D
(Yamaguchi and Mosher, 1973)
H21C10
CHCH=CH2
HO
(Ohta and Tetsukawa, 1980)
Me
Me
CDCHNH2
Me
(Kirmse and Krause, 1975)
Ar CH2CHNH2
Me
(Nicholas et al., 1973)
R—CHCH2OH
NH2
(Meyers et al., 1978;
Poindexter and Meyers, 1977)

Scheme 2. Some compounds for which configurations or enantiomeric compositions have been determined using MTPA and ^{19}F-NMR spectroscopy.

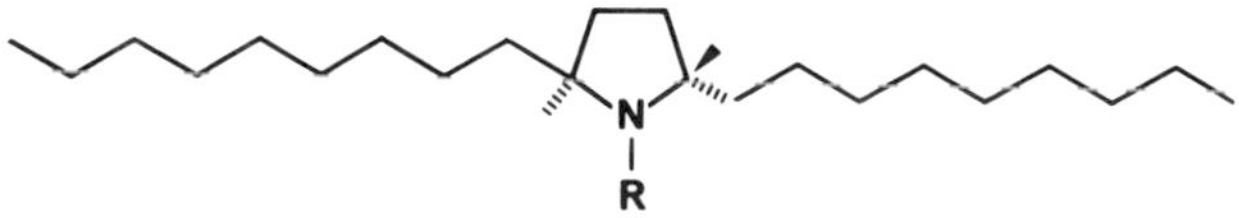
N
R

26A

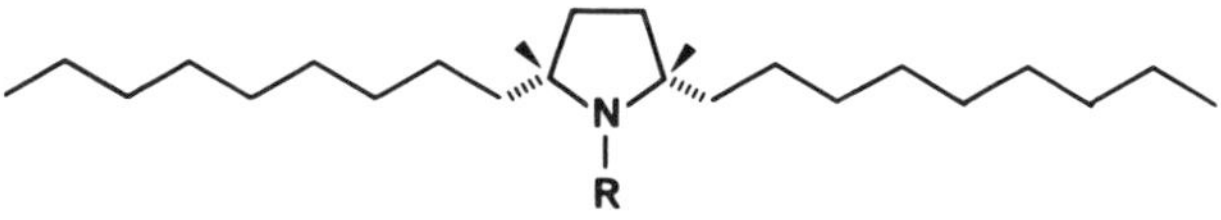
N
R

26B R= (R)—OMTPA

C. Miscellaneous Applications of Chemical Shift Nonequivalence

Chemical shift nonequivalence has been used to determine the configuration at the phosphorus center in diastereomeric l-menthyl esters of alkylphenyl phosphinates (27) (Lewis et al., 1968). The signal due to the (pro-S)-methyl of the menthyl isopropyl group in the epimers of 27 with the S-configuration at phosphorus appears consistently at higher field than that of the corresponding (pro-R)-methyl group. The enantiomeric purities of partially active phosphines (28) have been determined (Casey et al., 1969) by the conversion of these compounds to a mixture of diastereomeric phosphonium salts (30) with enantiomerically pure bromide 29. The ^{1}H-NMR spectra of diastereomeric osmate ester complexes (31), formed from osmate esters of glycols and optically active trans-N,N,N',N'-tetramethyl-1,2-cyclohexanediamine, have been used to determine the enantiomeric purity of the glycols (Resch and Meinwald, 1981).

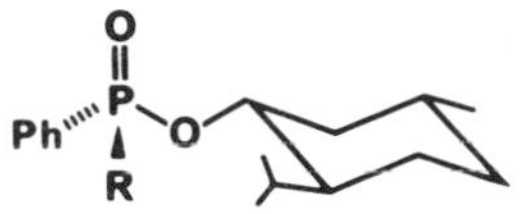

27

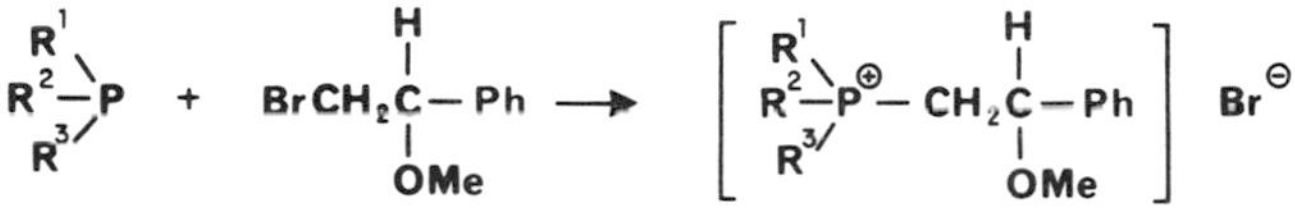

28   29   30

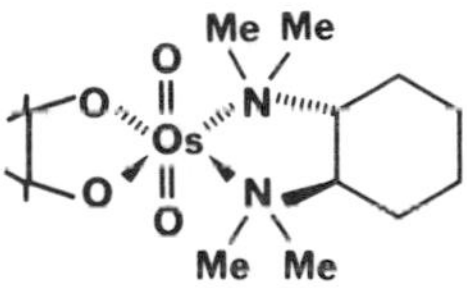

31

^{2}H-NMR has been used to determine the enantiomeric ratio of α-deuterated carboxylic acids and amines (Brown and Parker, 1981). A partially active substrate was converted to a diastereomeric pair of esters (32) with enantiomerically pure methyl mandelate; the enantiomeric purity was then determined using ^{2}H-NMR with broad-band proton decoupling.

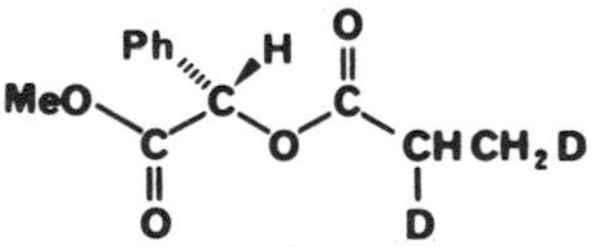

32

TABLE III
^{13}C-NMR Chemical Shift Differences[a] for C-1 to C-7
in Diastereomers 33 and 34[b]

		Atom						
Compound	X	1	2	3	4	5	6	7
33	O	0.00	0.96	0.50	0.00	0.38	1.07	0.11
34	S	0.08	1.27	0.77	0.00	0.71	1.33	0.12

[a]$\Delta\delta$ expressed in parts per million.
[b]From Hiemstra and Wynberg (1977).

^{13}C-NMR has been reported to be a useful technique for determining the enantiomeric composition of various cyclic and acyclic ketones (Helder *et al.*, 1977; Hiemstra and Wynberg, 1977). In ^{13}C-pulse NMR, signal intensities are not always proportional to the number of nuclei because of differences in relaxation time and Nuclear Overhauser effects. However, in cases in which the compounds have similar structures (e.g., when they are diastereomers), differences in these factors often may be negligible for the diastereotopic carbon atoms being compared. The signals with substantial chemical shift nonequivalence due to the diastereomeric carbon atoms were observed for cyclic ketals (33) and thioketals (34) derived from the corresponding ketones and enantiomerically pure 1,2-butanediol or 1,2-butanedithiol (Table III). The relative signal intensities indicated the enantiomeric composition of the original ketones. It is interesting that the largest chemical shift differences were not at the chiral center C-3 or the diastereotopic methyl groups at C-7, but at C-2 and C-6. A similar application of ^{13}C-NMR spectroscopy to the determination of the enantiomeric purity of 2-substituted cyclohexanones (Meyers *et al.*, 1981) and hydroxycarboxylic acids (Van Os *et al.*, 1979) has been reported.

The absolute configuration of secondary carbinols has been assigned using a glycosidation shift rule and ^{13}C-NMR spectroscopy (Seo *et al.*, 1978). The ^{13}C-NMR spectra of secondary alcoholic glucopyranosides were compared with those of the methyl glucoside and the corresponding parent alcohol to obtain the glycosidation shifts, $\Delta\delta S = \delta$(alcoholic glucoside) $- \delta$(methyl glucoside) for a sugar moiety and $\Delta\delta A = \delta$(alcoholic glucoside) $- \delta$(alcohol) for an aglycone moiety. These values have been correlated with the absolute configuration of the original carbinol. This method possesses wide applicability for determining the absolute stereochemistry of various secondary carbinols in aliphatic chains, five-membered rings, flexible medium-sized rings, and macrorings.

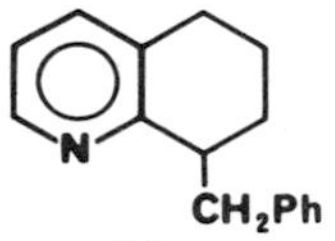

35

The interaction of salts from racemic 8-benzyl-5,6,7,8-tetrahydroquinoline (35) and optically active acids with the chiral complexing agent β-cyclodextrin results in a substantial diastereotopic splitting in ^{15}N-NMR spectra (Dyllick-Brenzinger and Roberts, 1980). This chiral recognition phenomenon might form the basis for a method of determining the absolute configuration of nitrogen bases or proton donors.

III. Application of Induced Chemical Shift Nonequivalence

Although the Mosher reagent is extremely useful, sometimes serious limitations are encountered. For example, a signal from an internally diastereotopic group of the diastereomeric pair may not separate sufficiently to afford reliable results, or the crucial signals in ^{1}H- and ^{19}F-NMR spectra may be complicated or overlap other resonances. These problems can be overcome in some cases by the use of more powerful NMR instruments, but if these are not readily available a lanthanide shift reagent might provide a solution.

A. Proton-Induced Chemical Shift Nonequivalence

It has been reported (Yamaguchi et al., 1972; Yamaguchi and Mosher, 1973) that diastereomeric mixtures of (R)-MTPA esters of partially active secondary carbinols (36), including α-d-benzyl alcohol, exhibit a dramatic lanthanide-in-

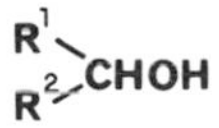

36

	A	B	C	D
R^1 =	nBu	Ph	Ph	Ph
R^2 =	tBu	Et	iPr	D

duced chemical shift (LIS) nonequivalence in ^{1}H-NMR signals from the OMe group in the MTPA moiety and the carbinyl proton in the alcohol moiety in the presence of the achiral lanthanide shift reagent Eu(fod)$_3$ [tris(6,6,7,7,8,8,8-heptafluoro-2,2-dimethyl 3,5-octanedionate)europium]. The enantiomeric ratio of the original carbinols can be determined from the relative intensities of the well-separated OMe or carbinyl diastereotopic proton signals.

Diastereomeric mixtures of (R)-MTPA esters of partially active methylethylcarbinol (**37**) [(S)-(+) excess, 61% ee] do not show any appreciable chemical shift nonequivalence for OMe and CF$_3$ groups in ^{1}H- and ^{19}F-NMR spectra (Fig. 4A), and partially overlapping signals were observed for the methyl and ethyl groups. Progressive addition of Eu(fod)$_3$ to the mixture, however, results in a substantial separation of OMe and other signals (Fig. 4B). The magnitude of the LIS for the OMe group in the MTPA moiety for the R,R-pair is larger than that for the R,S-isomer. Similarly, the LIS values for the ortho protons of the phenyl group and the methine proton (carbinyl moiety) in the R,R-pair are larger than those of the R,S-isomer. Extensive LIS studies have revealed that this regularity applies to more than 40 acyclic and cyclic carbinols; the diastereomeric (R)-MTPA ester with the larger LIS$_{OMe}$ value has configuration **38A**, whereas the alternate diastereomer with the smaller LIS$_{OMe}$ value has configuration **38B** (Table IV). This method is also adaptable to the case in which functional groups

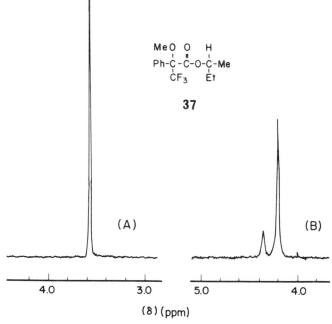

Fig. 4. 200-MHz Fourier transform ^{1}H-NMR spectra of OMe group in the (R)-MTPA esters of methylethylcarbinol [**37**; (S); 61% ee]; 1.0 mg in CCl$_4$ (A) in the absence of Eu(fod)$_3$ and (B) in the presence of 0.1 molar ratio of Eu(fod)$_3$ (no. of accumulation, 32).

TABLE IV

Chemical Shift Nonequivalences of the Methoxyl Group
in the Acid Moiety for Diastereomeric (R)-MTPA Esters
(38A and 38B) Induced by Eu(fod)$_3$[a]

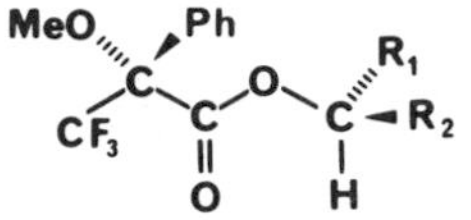

38A R$_1$ = R$_L$; R$_2$ = R$_M$

38B R$_1$ = R$_M$; R$_2$ = R$_L$

Ligand		Induced chemical shift difference $\Delta\delta$ (ppm),
R^1	R^2	LIS$_{OMe}$ (38A − 38B)
Et	Me	+2.7
tBu	Me	+7.5
Ph	Me	+4.0
Ph	tBu	+1.4
Ph	CH$_2$OMe	+2.2
Ph	CH$_2$CH$_2$CH$_2$SMe	+0.8
Menthol		+7.7
Borneol		+0.2
3-α-Cholestanol		+4.1

[a]Data from Yamaguchi and Kabuto (1977) and Yamaguchi et al. (1976).

such as OMe, SMe, or NMe$_2$ are in the carbinyl moiety (Yamaguchi et al., 1976; Yamaguchi and Kabuto, 1977), allowing simultaneous determination of configuration and cc. The presence of an additional chiral center (or centers) in the substrate does not alter the reliability of the method because the effect due to remote chiral centers is relatively small. For example, ΔLIS$_{OMe}$ (R,R − R,S) for **39**, which has a β-carbinyl chiral carbon, is small (Yasuhara and Yamaguchi, 1977) compared with the value for **40** (Yamaguchi et al., 1976).

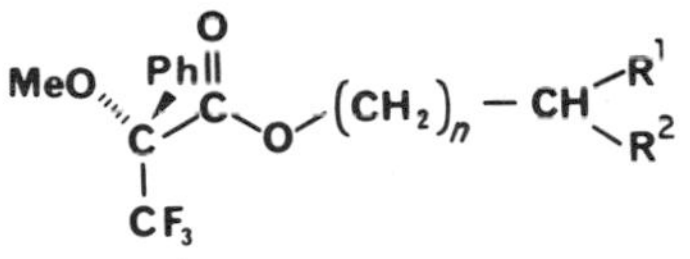

39 n = 1

40 n = 0

An empirical model that rationalizes the observed NMR configurational correlation has been proposed (Fig. 5). The essential role of Eu(fod)$_3$ may be that of

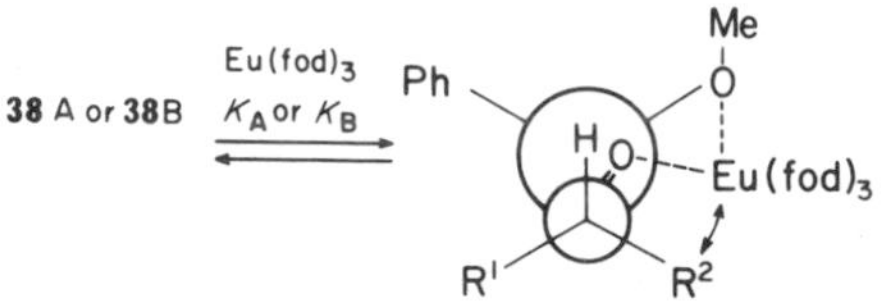

Fig. 5. Conformational correlation model for (*R*)-MTPA and (*S*)-MTPA derivatives **38A** and **38B**, respectively, in the presence of Eu(fod)$_3$ (**38A**, R^1 = L, R^2 = M; **38B**, R^1 = M, R^2 = L; $k_A > k_B$; LIS$_{OMe}^A$) > LIS$_{OMe}^B$). From Yamaguchi *et al.*, (1976).

approaching the diastereotopic face of the ester carbonyl group of the MTPA ester, differentiating the chirality and also the steric requirement of the carbinyl moiety. At the same time it could reduce the conformational freedom of the MTPA moiety by coordinating the oxygen atoms of both the ester carbonyl and the OMe group.

A conformational change in the substrate upon coordination of Eu(fod)$_3$ has been reported (Hofer, 1976). A Job plot (Roth *et al.*, 1972; Hofer, 1976), on **41** indicates that the MTPA ester forms a 1 : 1 complex (F. Yasuhara, K. Kabuto, and S. Yamaguchi, unpublished results), with Eu(fod)$_3$ acting as a bidenate ligand. The magnitude of LIS$_{OMe}$ for **41** (9.6 ppm) is considerably larger than that for **42** (1.6 ppm), in which the ester carbonyl group cannot participate in the

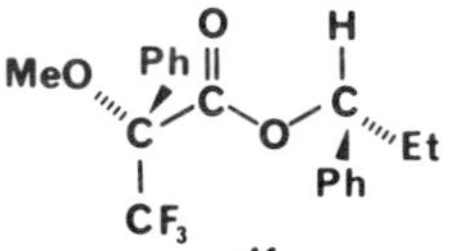

formation of a five-membered chelate ring. Along with previous findings (Yamaguchi *et al.*, 1976), these observations (K. Kabuto and S. Yamaguchi, unpublished results), are consistent with the proposed model. This NMR configurational correlation model is not reliable when the effective steric bulk differences between R_L and R_M (L = large-, M = medium-size groups) are substantially smaller (Yamaguchi *et al.*, 1976) or when the substrate possesses several effective Eu(fod)$_3$ coordination sites [e.g., Lys(Ts)—OMe or DOPA(Me)$_2$] (K. Kabuto and S. Yamaguchi, unpublished results). Furthermore, although the reason remains equivocal, certain cyclic compounds, for example, *cis*-3-methyl- (Yamaguchi *et al.*, 1976) and *cis*-3-*tert*-butylcyclohexanols (Lightner *et al.*, 1981), do not follow the correlation model. A warning about the use of a single conformation to explain LIS has been discussed (Sullivan, 1976).

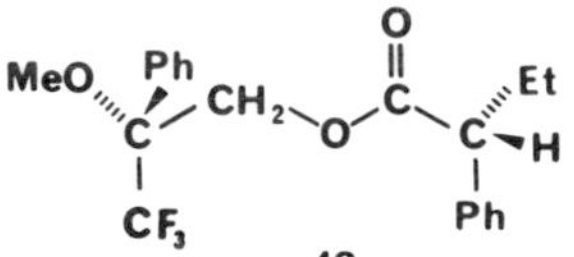

Application of this MTPA–lanthanide shift reagent method has been extended (Yamaguchi and Yasuhara, 1977) to epimeric mixtures of secondary carbinols. It is interesting that the sign of ΔLIS$_{OMe}$ for the diastereomeric MTPA esters (e.g.,

43 and **44**) is governed solely by the absolute configuration of the carbinyl carbon atom in question. Thus, the configuration of carbinols in epimeric mixtures can be determined independently without any stereochemical information about the neighboring chiral center. These observations may be interpreted as follows. The major conformations of the MTPA esters of *l*-menthol and *d*-neomenthol that coordinate with Eu(fod)$_3$ are represented by **43** and **44**, respectively. The absolute stereochemistry at the carbinyl carbon atom determines the position of the *i*Pr group, that is, on which diastereoface of the ester carbonyl group is resides—toward OMe or away from it. As discussed above, the steric effect of the *i*Pr group in these MTPA esters substantially influences the relative ease of complex formation with Eu(fod)$_3$ and thus affects the relative magnitude of the LIS$_{OMe}$ value (LIS$_{OMe}$ for **43** and **44** is 12.4 and 9.0 ppm, respectively). In contrast, stereochemical features based on more remote relationships, such as *cis trans* and *axial–equatorial* orientations of the hydroxyl function, do not alter substantially the relative ease of complex formation, and thus the relative sense of the induced chemical shift nonequivalence is not affected.

Absolute stereochemistry and enantiomeric ratios of various types of compounds such as bicyclic secondary carbinols (Kalyanam and Lightner, 1979; Lightner *et al.*, 1980), acyclic primary carbinols (Yasuhara and Yamaguchi, 1977), substituted cyclopropanemethanols (F. Yasuhara, K. Kabuto, and S. Yamaguchi, unpublished results) with the chiral center at the C-2 position, α- and β-amino acid esters (Yasuhara *et al.*, 1978), and α- and β-hydroxy acid esters (Yasuhara and Yamaguchi, 1980) have been determined by the MTPA–lanthanide shift reagent method. The enantiomeric purity of substituted cyclopentane- and cyclohexanemethanols in both mono- and bicyclic modifications has also been determined (F. Yasuhara and S. Yamaguchi, unpublished results). Structures representing other examples are shown in Scheme 3.

The application of the MTPA–lanthanide shift reagent method to various types of biaryls and bridged biaryls that carry hydroxyl, amino, carboxyl, and hydroxymethylene functionalities yielded the first examples of simultaneous determination by NMR of absolute configuration and enantiomeric purity (Kabuto *et al.*, 1980, 1981) for compounds with axial chirality. The maximum optical rotation of 2-hydroxybinaphthyl (**45**), for which quite disparate values have been reported (Jacques and Fouquey, 1971; Helgeson *et al.*, 1974; Grubbs and DeVries, 1977), could be determined (Kabuto *et al.*, 1980; Miyano *et al.*, 1980, 1981) accurately by this method. MTPA–lanthanide shift reagent and chemical

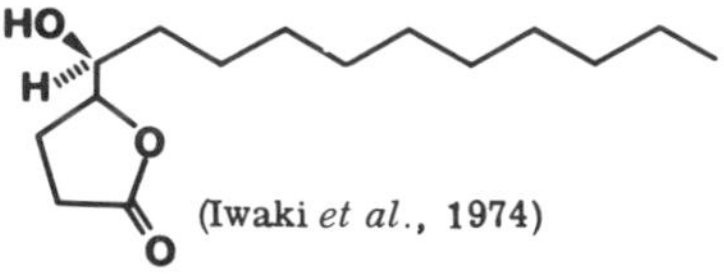

(Iwaki *et al.*, 1974)

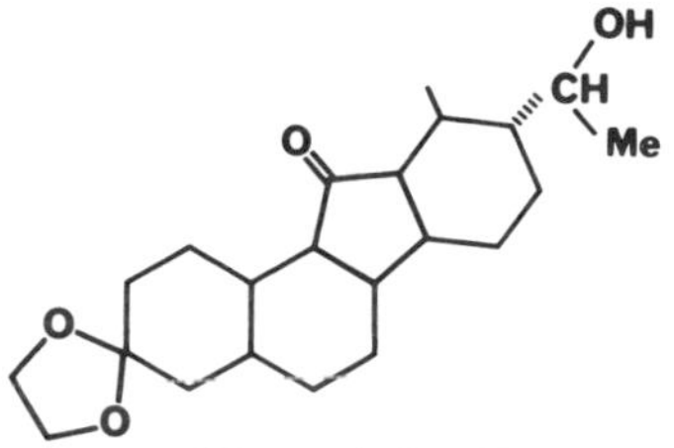

(Murai *et al.*, 1977)

D
|
PhCHOH

(Yamaguchi and Mosher, 1973)

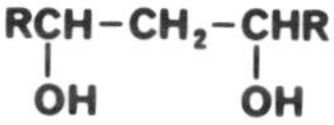

RCH–CH₂–CHR
| |
OH   OH

(F. Yasuhara and
S. Yamaguchi, unpubl.)

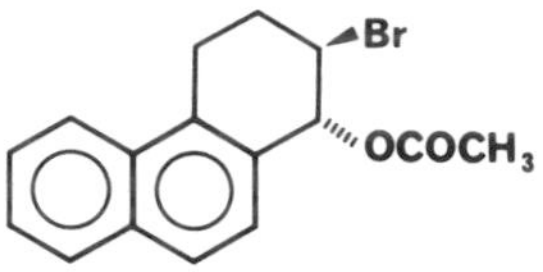

(Boyd *et al.*, 1977)

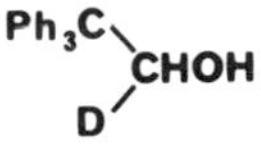

Ph₃C
  \
   CHOH
  /
 D

(Reich *et al.*, 1973)

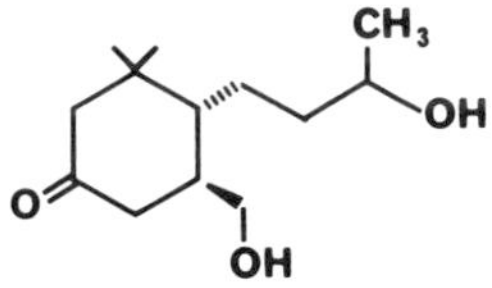

(Sugimoto *et al.*, 1978, 1979)

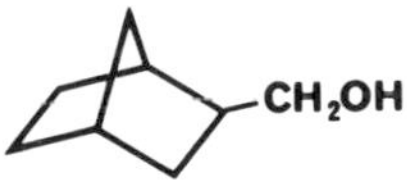

(F. Yasuhara and
Y. Yamaguchi, unpubl.)

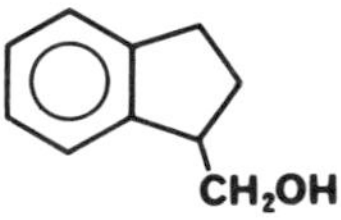

(F. Yasuhara, K. Kabuto, and
S. Yamaguchi, unpubl.)

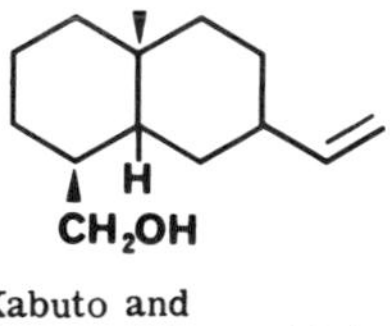

(K. Kabuto and
S. Yamaguchi, unpubl.)

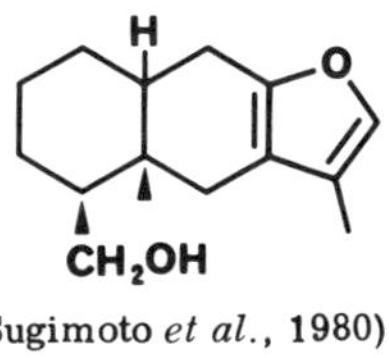

(Sugimoto *et al.*, 1980)

Scheme 3. Some compounds for which configurations or enantiomeric compositions have been determined using MTPA and the lanthanide shift reagent method.

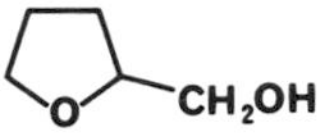

O
CH₂OH

(F. Yasuhara and S. Yamaguchi, unpubl.)

CH₂OH

(F. Yasuhara and S. Yamaguchi, unpubl.)

S
S
OH

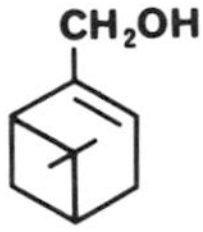

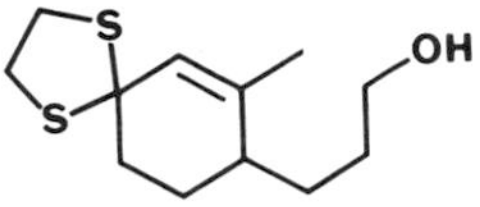

(Johnson *et al.*, 1980)

OH
CH
Me
Et

(F. Yasuhara and S. Yamaguchi, unpubl.

OH

(Cambie *et al.*, 1975)

OH
C₅H₁₁CHCH=CH₂

(Oritani and Yamashita, 1980)

OH
C₅H₁₁CHCH=CHCH=CH(CH₂)₇COOMe

(Van Os *et al.*, 1979)

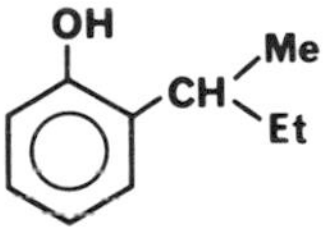

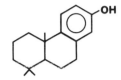

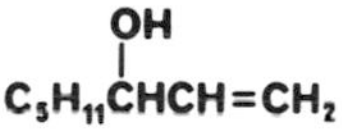

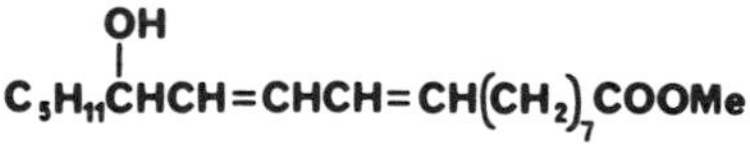

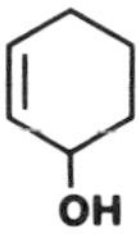

OH

(F. Yasuhara and S. Yamaguchi, unpubl.)

OH

(Mori *et al.*, 1978)

OH
n-C₅H₁₁CHC≡CH

(Oritani and Yamashita, 1980)

R—CHC≡C—R'
OH

(Mori and Akao, 1978)

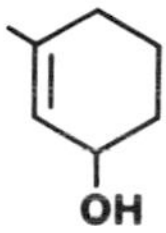

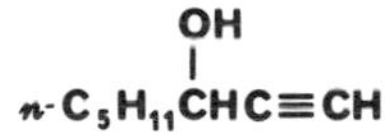

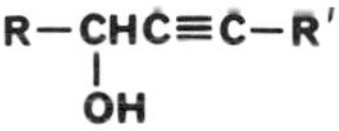

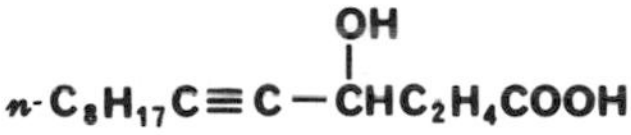

OH
n-C₈H₁₇C≡C—CHC₂H₄COOH

(Nishizawa *et al.*, 1981)

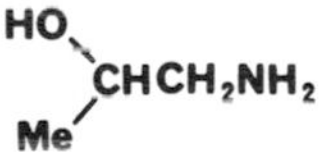

HO
CHCH₂NH₂
Me

(F. Yasuhara and S. Yamaguchi, unpubl.; via MTPA oxazoline derivative)

HO— CH₂CH
Ph
NH₂

(F. Yasuhara and S. Yamaguchi, unpubl.; via MTPA oxazoline derivative)

CH₂OH

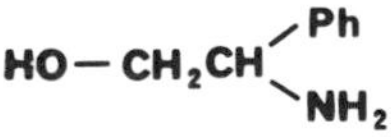

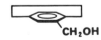

(K. Kabuto and S. Yamaguchi, unpubl.)

Scheme 3 (*Continued*)

correlation studies on **45** have also revealed that the previous assignment, (R)-$(-)$ in tetrahydrofuran (THF) (Berson and Greenbaum, 1958; Akimoto and Yamada, 1971), should be revised (R)-$(+)$ in THF and (R)-$(-)$ in MeOH (Kabuto et al., 1981). Empirical models that rationalize the observed chemical shift nonequivalence for the MTPA derivatives of these compounds with axial chirality have been proposed. Simultaneous determination of enantiomeric ratio and absolute stereochemistry of various cis- and trans-hydroxyketo and dihydroxyspiro compounds with C_2 axes represents a useful application (K. Kabuto, F. Yasuhara, and S. Yamaguchi, unpublished results) of the MTPA–lanthanide shift reagent method.

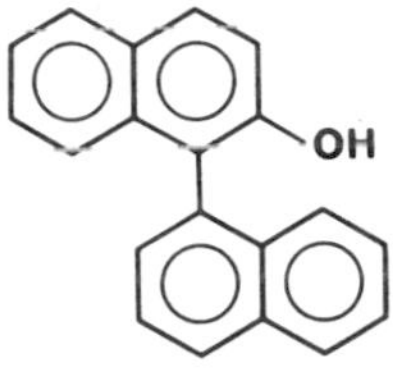

45 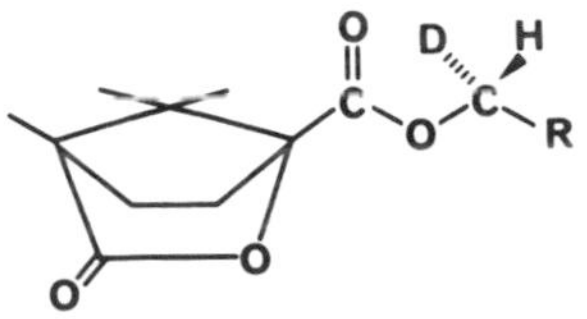   46

Camphanic acid is a valuable chiral derivatizing agent for use in LIS–^{1}H-NMR studies of α-deuterated primary carbinols (Gerlach and Zagalak, 1973). In these esters (**46**) the signal due to the (pro-R)-carbinyl hydrogen appears consistently at higher field than that due to the (pro-S)-hydrogen, thus allowing configurational assignments to be made for these carbinols.

The enantiomeric ratio of 7-methoxy-5,6-dimethyl-1,2,3,4-tetrahydro-1-naphthoic acid (**47**) (Feringa and Wynberg, 1981) and the absolute configuration

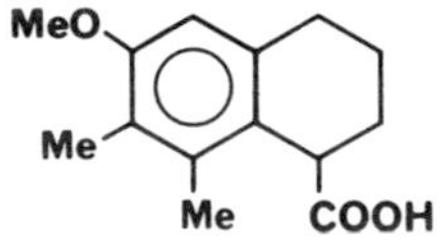

47 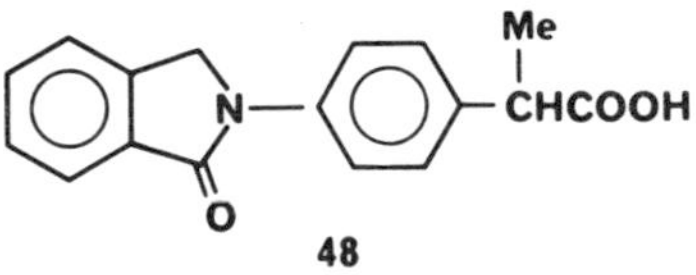   48

of 2-[4-(1-oxo-2-isoindolinyl)phenyl]propionic acid (**48**) (Munari et al., 1980) have been determined by the use of their diastereomeric phenylethylamine amides in the presence of $Eu(fod)_3$.

B. Fluorine-Induced Chemical Shift Nonequivalence

Lanthanide-induced shift studies with ^{19}F-NMR spectroscopy have proved useful for determining the enantiomeric purity and absolute configuration of substrates. Detailed investigations with various bicyclic secondary carbinols (Kalyaman and Lightner, 1979) and cyclohexanols (Merckx et al., 1978; Van de Waal et al., 1978) have revealed that an NMR configurational correlation scheme which is the same as that developed in ^{1}H-NMR spectroscopy (Yamaguchi *et al.*, 1976) holds well for these types of compounds. A similar application of the LIS method in ^{19}F-NMR spectroscopy has been reported for *endo*-norbornenylmethanol (Oppolzer *et al.*, 1981) and hydroxy acid ester (Van Os *et al.*, 1979).

Acknowledgment

I would like to express my sincere thanks to Professor Harry S. Mosher for providing many valuable references and exceedingly helpful comments.

References

Ackermann, P., Tabler, H., and Ganter, C. (1972). *Helv. Chim. Acta* **55**, 2731.

Afzali-Ardakani, A., and Rapoport, H. (1980). *J. Org. Chem.* **45**, 4817.

Akimoto, H., and Yamada, S. (1971). *Tetrahedron* **27**, 5999.

Albers-Schönberg, G., Arison, B. H., Hensens, O. D., Hirshfield, J., Hoogsteen, K., Kaczka, E. A., Rhodes, R. E., Kahan, J. J., Kahan, F. M., Ratcliffe, R. W., Walton, E., Ruswinkle, L. J., Morin, R. B., and Christensen, B. G. (1978). *J. Am. Chem. Soc.* **100**, 6491.

Balani, S. K., Boyd, D. R., Cassidy, E. S., Greene, R. M. E., McCombe, K. M., and Sharma, N. D. (1981). *Tetrahedron Lett.* **22**, 3277.

Barth, G., Voeter, W., Mosher, H. S., Bunnenberg, E., and Djerassi, C. (1970). *J. Am. Chem. Soc.* **92**, 875.

Baxter, C. A. R., and Richards, H. C. (1972). *Tetrahedron Lett.*, p. 3357.

Becker, Y., Eisenstadt, A., and Stille, J. K. (1980). *J. Org. Chem.* **45**, 2145.

Bernasconi, S., Corbella, A., Gariboldi, P., and Jommi, G. (1977). *Gazz. Chim. Ital.* **107**, 95.

Berson, J. A., and Greenbaum, M. A. (1958). *J. Am. Chem. Soc.* **80**, 653.

Boyd, D. R., Neill, J. D., and Stubbs, M. E. (1977). *J. Chem. Soc. Chem. Commun.*, p. 873.

Brown, J. M., and Parker, D. (1981). *Tetrahedron Lett.* **22**, 2815.

Cambie, R. C., David, B. R., Hayward, R. C., and Woodgate, P. D. (1975). *Aust. J. Chem.* **28**, 631.

Casey, J. P., Lewis, R. A., and Mislow, K. (1969). *J. Am. Chem. Soc.* **91**, 2789.

Cochran, T. G., and Huitric, A. C. (1971). *J. Org. Chem.* **36**, 3043.

Cohen, N., Lopresti, R. J., Neukom, C., and Saucy, G. (1980). *J. Org. Chem.* **45**, 582.

Dale, J. A., and Mosher, H. S. (1968). *J. Am. Chem. Soc.* **90**, 3732.

Dale, J. A., and Mosher, H. S. (1973). *J. Am. Chem. Soc.* **95**, 512.

Dale, J. A., Dull, D. L., and Mosher, H. S. (1969). *J. Org. Chem.* **34**, 2543.

Dutcher, J. S., MacMillan, J. G., and Heathcock, C. H. (1976). *J. Org. Chem.* **41**, 2663.

Dyllick-Brenzinger, R., and Roberts, J. D. (1980). *J. Am. Chem. Soc.* **102**, 1166.

Eliel, E. L., and Lynch, J. E. (1981). *Tetrahedron Lett.* **22**, 2855.

Feigl, D. M. (1965). Ph.D. Thesis, Stanford Univ., Stanford, California.

Felsky, G., Fredericks, P. M., and Meakins, G. D. (1978). *J. Chem. Soc. Perkin Trans. 1*, p. 1529.

Feringa, B., and Wynberg, H. (1981). *J. Org. Chem.* **46**, 2547.

Fraser, R. R., Pett, M. A., and Saunders, J. K. (1971). *J. Chem. Soc. Chem. Commun.*, p. 1450.

Galpin, D. R., and Huitric, A. C. (1968). *J. Org. Chem.* **33**, 921.

Gaudemer, A. (1977). *In* "Stereochemistry" (H. B. Kagan, ed.), Vol. I, p. 117. Thieme, Stuttgart.

Gerlach, H. (1966). *Helv. Chim. Acta* **49**, 2481.

Gerlach, H., and Zagalak, B. (1973). *J. Chem. Soc. Chem. Commun.* p. 274.

Goering, H. L., Eikenberry, J. N., Koermer, G. S., and Lattimer, C. J. (1974). *J. Am. Chem. Soc.* **96**, 1493.

Grubbs, R. H., and DeVries, R. A. (1977). *Tetrahedron Lett.*, p. 1879.

Hayashi, H., and Nakanishi, K. (1973). *J. Am. Chem. Soc.* **95**, 4081.

Hayashi, H., Nakanishi, K., Brandon, C., and Marmur, J. (1973). *J. Am. Chem. Soc.* **95**, 8749.

Helder, R., Arends, R., Bolt, H., Hiemstra, H., and Wynberg, H. (1977). *Tetrahedron Lett.*, p. 2181.

Helgeson, R. C., Timko, J. M., Moreau, P., Peacock, S. C., Mayer, J. M., and Cram, D. J. (1974). *J. Am. Chem. Soc.* **96**, 6762.

Helmchen, G. (1974). *Tetrahedron Lett.*, p. 1527.

Helmchen, G., and Schmierer, R. (1976). *Angew. Chem. Int. Ed. Engl.* **15**, 703.

Helmchen, G., Roh, R., and Sanber, K. (1972). *Tetrahedron Lett.*, p. 3873.

Helmchen, G., Völter, H., and Schüle, W. (1977). *Tetrahedron Lett.*, p. 1417.

Helmchen, G., Nill, G., Flockerzi, D., Schühle, W., and Youssef, M. S. K. (1979a). *Angew. Chem. Int. Ed. Engl.* **18**, 62.

Helmchen, G., Nill, G., Flockerzi, D., and Youssef, M. S. K. (1979b). *Angew. Chem. Int. Ed. Engl.* **18**, 63.

Herold, T., and Hoffmann, R. W. (1978). *Angw. Chem. Int. Ed. Engl.* **17**, 768.

Hiemstra, H., and Wynberg, H. (1977). *Tetrahedron Lett.*, p. 2183.

Hofer, O. (1976). *Top. Stereochem.* **9**, 111.

Horeau, A. (1969). *Tetrahedron Lett.*, p. 3121.

Horn, D. H. S., and Pterorius, Y. Y. (1954). *J. Chem. Soc.*, p. 1460.

Hoyer, G. A., Rosenberg, D., Rufer, C., and Seeger, A. (1972). *Tetrahedron Lett.*, p. 985.

Ishizaki, T., Miura, H., and Nohira, H. (1980). *Nippon Kagaku Zasshi* **8**, 1381.

Iwaki, S., Marumo, S., Saito, T., Yamada, M., and Katagiri, K. (1974). *J. Am. Chem. Soc.* **96**, 7842.

Jacobus, J., Raban, M., and Mislow, K. (1968). *J. Org. Chem.* **33**, 1142.

Jacques, J., and Fouquey, C. (1971). *Tetrahedron Lett.*, p. 4617.

Johnson, W. S., McCarry, B. E., Markezich, R. L., and Boots, S. G. (1980). *J. Am. Chem. Soc.* **102**, 352.

Kabuto, K., Yasuhara, F., and Yamaguchi, S. (1980). *Tetrahedron Lett.* **21**, 307.

Kabuto, K., Yasuhara, F., and Yamaguchi, S. (1981). *Tetrahedron Lett.* **22**, 659.

Kalyanam, N., and Lightner, D. A. (1979). *Tetrahedron Lett.*, p. 415.

Kawana, H., and Ishikawa, N. (1980). *Chem. Lett.*, p. 843.

Kirmse, W., and Krause, D. (1975). *Chem. Ber.* **108**, 1855.

Klyne, W., and Buckingham, J. (1978). "Atlas of Stereochemistry," 2nd ed., Vol. 1. Chapman & Hall, London.

Lee, T. D., and Keana, F. W. (1978). *J. Org. Chem.* **43**, 4226.

Lewis, R. A., Korpium, O., and Mislow, K. (1968). *J. Am. Chem. Soc.* **90**, 4847.

Lightner, D. A., Gawronski, J. K., and Bouman, T. D. (1980). *J. Am. Chem. Soc.* **102**, 1983.

Lightner, D. A., Bowman, T. D., Gawronski, J. K., Gawronska, K., and Chappuis, J. L. (1981). *J. Am. Chem. Soc.* **103**, 5314.

Martin, V. S., Woodard, S. S., Katsuki, T., Yamada, Y., Ikeda, M., and Sharpless, K. B. (1981). *J. Am. Chem. Soc.* **103**, 6237.

Mathieson, A. M. (1965). *Tetrahedron Lett.*, p. 4137.

Mazaleyrat, J., and Cram, D. J. (1981). *J. Am. Chem. Soc.* **103**, 4585.

McCreary, M. D., Lewis, D. W., Wernick, D. L., and Whitesides, G. M. (1974). *J. Am. Chem. Soc.* **96**, 1038.

McGahren, W. J., Sax, K. J., Kunstmann, M. P., and Ellestad, G. A. (1977). *J. Org. Chem.* **42**, 1659.

Merckx, E. M., Van de Waal, A. J., Lepoivre, J. A., and Alderweireldt, F. C. (1978). *Bull. Soc. Chim. Belg.* **87**, 21.

Meyers, A. I., and Birch, Z. (1979). *J. Chem. Soc. Chem. Commun.*, p. 567.

Meyers, A. I., and Hudspeth, J. P. (1981). *Tetrahedron Lett.* **22**, 3925.

Meyers, A. I., Poindexter, G. S., and Birch, Z. (1978). *J. Org. Chem.* **43**, 892.

Meyers, A. I., Williams, D. R., Erickson, G. W., White, S., and Druelinger, M. (1981). *J. Am. Chem. Soc.* **103**, 3081.

Miyano, S., Tobita, M., Suzuki, S., Nishikawa, Y., and Hashimoto, H. (1980). *Chem. Lett.*, p. 1027.

Miyano, S., Tobita, M., and Hashimoto, H. (1981). *Bull. Chem. Soc. Jpn.* **54**, 3522.

Mori, K. (1975). *Tetrahedron* **31**, 3011.

Mori, K. (1976). *Agric. Biol. Chem.* **40**, 415.

Mori, K., and Akao, H. (1978). *Tetrahedron Lett.*, p. 4127.

Mori, K., Mizumachi, N., and Matsui, M. (1976a). *Agric. Biol. Chem.* **40**, 1611.

Mori, K., Takigawa, T., and Matsui, M. (1976b). *Tetrahedron Lett.*, p. 3953.

Mori, K., Tamada, S., Uchida, M., Mizumachi, N., Tachibana, Y., and Matsui, M. (1978). *Tetrahedron* **34**, 1901.

Mukaiyama, T., and Suzuki, K. (1980). *Chem. Lett.*, p. 255.

Mukaiyama, T., Sakito, Y., and Asami, M. (1978). *Chem. Lett.*, p. 1253.

Mukaiyama, T., Suzuki, K., Soai, K., and Sato, T. (1979). *Chem. Lett.*, p. 447.

Munari, S. D., Marazzi, G., Forgione, A., Longo, A., and Lombardi, P. (1980). *Tetrahedron Lett.* **21**, 2273.

Murai, A., Sasamori, H., and Masamune, T. (1977). *Chem. Lett.*, p. 669.

Nelson, W. L., and Burke, T. R., Jr (1978). *J. Org. Chem.* **43**, 3641.

Nicholas, D. E., Bartknecht, C. F., Rusterholz, D. B., Benington F., and Morin, R. D. (1973). *J. Med. Chem.* **16**, 480.

Nishizawa, M., Yamada, M., and Noyori, R. (1981). *Tetrahedron Lett.* **22**, 247.

Noyori, R., Tomino, I., and Tanimoto, Y. (1979). *J. Am. Chem. Soc.* **101**, 3129.

Ohta, H., and Tetsukawa, H. (1980). *Agric. Biol. Chem.* **44**, 863.

Oppolzer, W., Kurth, M., Reichlin, D., and Moffatt, F. (1981). *Tetrahedron Lett.* **22**, 2545.

Oritani, T., and Yamashita, K. (1980). *Agric. Biol. Chem.* **44**, 2407.

Partridge, J. J., Chadha, N. K., and Uskoković, M. R. (1973). *J. Am. Chem. Soc.* **95**, 532.

Pirkle, W. H., and Hauske, J. R. (1976). *J. Org. Chem.* **41**, 801.

Pirkle, W. H., and Sikkenga, D. L. (1977). *J. Org. Chem.* **42**, 1370.

Pirkle, W. H., and Simmons, K. A. (1979). *J. Org. Chem.* **44**, 4891.

Pirkle, W. H., and Simmons, K. A. (1981). *J. Org. Chem.* **46**, 3239.

Pohl, L. R., and Trager, W. F. (1973). *J. Med. Chem.* **16**, 475.

Poindexter, G. S., and Meyers, A. I. (1977). *Tetrahedron Lett.*, p. 3527.

Raban, M., and Mislow, K. (1967). *Top. Stereochem.* **2**, 199.

Reich, C. J. (1976). Ph.D. Thesis, Stanford Univ., Stanford, California.

Reich, C. J., Sullivan, G. R., and Mosher, H. S. (1973). *Tetrahedron Lett.*, p. 1505.

Resch, J. F., and Meinwald, J. (1981). *Tetrahedron Lett.* **22**, 3159.

Roth, K., Grosse, M., and Rewicki, D. (1972). *Tetrahedron Lett.*, p. 435.

Sandman, D. J., Mislow, K., Giddings, W. P., Dirlam, J., and Hanson, G. C. (1968). *J. Am. Chem. Soc.* **90**, 4877.

Schlosser, M., and Fouquet, G. (1974). *Chem. Ber.* **107**, 1162.

Schurig, V., Koppenhöfer, B., and Bürkle, W. (1978). *Angew. Chem. Int. Ed. Engl.* **17**, 937.

Seo, S., Tomita, Y., Tori, K., and Yoshimura, Y. (1978). *J. Am. Chem. Soc.* **100**, 3331.

Soai, K., and Mukaiyama, T. (1979). *Bull. Chem. Soc. Jpn.* **52**, 3371.

Sugimoto, Y., Sakita, T., Moriyama, Y., Murae, T., Tsuyuki, T., and Takahashi, T. (1978). *Tetrahedron Lett.*, p. 4285.

Sugimoto, Y., Sakita, T., Ikeda, T., Moriyama, Y., Murae, T., Tsuyuki, T., and Takahishi, T. (1979). *Bull. Chem. Soc. Jpn.* **52**, 3027.

Sugimoto, Y., Tsuyuki, T., Moriyama, Y., and Takahashi, T. (1980). *Bull. Chem. Soc. Jpn.* **53**, 3723.

Sugiyama, T., and Yamashita, K. (1980). *Agric. Biol. Chem.* **44**, 1983.

Sullivan, G. R. (1976). *J. Am. Chem. Soc.* **98**, 7162.

Sullivan, G. R., Dale, J. A., and Mosher, H. S. (1973). *J. Org. Chem.* **38**, 2143.

Trost, B. M., and Curran, D. P. (1981). *Tetrahedron Lett.* **22**, 4929.

Trost, B. M., O'Krongly, D., and Belletire, J. L. (1980). *J. Am. Chem. Soc.* **102**, 7595.

Valente, E. J., Pohl, L. R., and Trager, W. F. (1980). *J. Org. Chem.* **45**, 543.

Van de Waal, A. J., Merckx, E. M., Lemiere, G. L., Van Osselaer, T. A., Lepoivre, J. A., and Alderweireldt, F. C. (1978). *Bull. Soc. Chim. Belg.* **87**, 545.

Van Os, C. P. A., Vente, M. and Vliegenthart, J. F. G. (1979). *Biochim. Biophys. Acta* **574**, 103.

Wüthrich, H. J., Siewinski, A., Schaffner, H., and Jeger, O. (1973). *Helv. Chim. Acta* **56**, 239.

Yamaguchi, S., and Kabuto, K. (1977). *Bull. Chem. Soc. Jpn.* **50**, 3074.

Yamaguchi, S., and Mosher, H. S. (1973). *J. Org. Chem.* **38**, 1870.

Yamaguchi, S., and Yasuhara, F. (1977). *Tetrahedron Lett.*, p. 89.

Yamaguchi, S., Dale, J. A., and Mosher, H. S. (1972). *J. Org. Chem.* **37**, 3174.

Yamaguchi, S., Yasuhara, F., and Kabuto, K. (1976). *Tetrahedron* **32**, 1363.

Yasuhara, F., and Yamaguchi, S. (1977). *Tetrahedron Lett.*, p. 4085.

Yasuhara, F., and Yamaguchi, S. (1980). *Tetrahedron Lett.* **21**, 2827.

Yasuhara, F., Kabuto, K., and Yamaguchi, S. (1978). *Tetrahedron Lett.*, p. 4289.

8

Nuclear Magnetic Resonance Analysis Using Chiral Solvating Agents

Gary R. Weisman
Department of Chemistry
University of New Hampshire
Durham, New Hampshire

I. Introduction

Mislow and Raban (1967) proposed and Pirkle (1966) experimentally demonstrated that chiral solute enantiomers exhibit different NMR spectra when dissolved in a nonracemic chiral solvent. Since that time, many examples of the use of chiral solvents and other diamagnetic *chiral solvating agents* (CSAs) (Pirkle and Hoover, 1982) for NMR studies of chiral solutes have appeared in the literature. The principal applications reported to date have been determinations of enantiomeric purity and simple demonstrations of chirality, determinations of absolute configuration, and determinations of the kinetics of enantiomerizations. One outdated review (Ejchart and Jurczak, 1970) and several noncomprehensive but useful reviews of the subject (Gaudemer, 1977; Martin *et al.*, 1980; Jacques *et al.*, 1981; Erdik, 1981) are available. An excellent comprehensive review of NMR applications of CSAs has appeared (Pirkle and Hoover, 1982). The review is especially authoritative in light of the fact that Pirkle and co-workers have carried out a substantial portion of the work in this area.

The purpose of this brief chapter is to acquaint the organic chemist with the principles and practice of using CSAs for the NMR determination of enantiomeric purity or for the demonstration of chirality. With an aim toward providing a concise and useful reference source and complementing the review of Pirkle and Hoover (1982), a compilation of reported CSA–NMR applications, categorized according to chiral solute functional group, is included. The application of CSAs to the determination of absolute configuration is not discussed herein; that topic has been detailed in the Pirkle and Hoover (1982) review.

II. Principles

Consider chiral solute enantiomers S_d and S_l containing corresponding sensor nuclei, the latter enantiotopic by external comparison (e.g., the enantiomers of alanine and their methine protons) (Mislow and Raban, 1967). In an achiral environment, enantiotopic nuclei are isochronous (chemical shift equivalent). That is, enantiomeric solutes and mixtures of these solutes including racemates, when dissolved in typical achiral NMR solvents, normally yield identical NMR spectra. However, enantiotopic nuclei become diastereotopic when placed in a chiral environment and can, in principle, become anisochronous (chemical shift nonequivalent).

When the chiral environment is a CSA, the observed anisochrony is related to the solvation of solute enantiomers by CSA and is most easily understood if this solvation is considered to consist of the formation of a binary association complex between enantiomerically pure CSA (C_d or C_l) and each of the solute enantiomers (S_d and S_l). Appropriate equilibrium expressions are given as Eqs. (1) and (2), where $S_d C_d$ and $S_l C_d$ are complexes. The assumption of binary

$$S_d + C_d \overset{K}{\rightleftharpoons} S_d C_d \tag{1}$$

$$S_l + C_d \overset{K'}{\rightleftharpoons} S_l C_d \tag{2}$$

association is well justified in many instances (see following discussion) but is probably an oversimplification in others. The basic principles, however, should hold for more complicated equilibria as well as for Eqs. (1) and (2).

Under conditions of fast exchange, two NMR signals due to the sensor nuclei may be observed, one corresponding to the weighted average of the two signals from S_d and S_dC_d and the other corresponding to the weighted average of the two signals from S_l and S_lC_d. Chemical shifts of these observed resonances, δ_{obs} and δ'_{obs}, can be expressed in terms of Eqs. (3) and (4), where δ_X terms correspond to the chemical shifts (in parts per million) of species X resonances, and p and p' are the fractional populations of unsolvated S_d and S_l, respectively (Martin et al., 1980, p. 392). Assuming that the sensor nuclei in unsolvated S_d and S_l are

$$\delta_{obs} = p\delta_{S_d} + (1 - p)\delta_{S_dC_d} \tag{3}$$

$$\delta'_{obs} = p'\delta_{S_l} + (1 - p')\delta_{S_lC_d} \tag{4}$$

$$\Delta\delta = |\delta_{obs} - \delta'_{obs}| \tag{5}$$

isochronous, that is, that $\delta_{Sd} = \delta_{Sl}$, the observed anisochrony $\Delta\delta$ [Eq. (5)] has two sources. The source most easily recognized is the intrinsic anisochrony between the sensor nuclei in instantaneous *diastereomeric* association complexes S_dC_d and S_lC_d. The other is the difference between the fractional populations of the association complexes, which is, of course, related to species concentrations and to the relative magnitudes of association equilibrium constants K and K'. It can be shown that the CSA need not be enantiomerically pure but only enantiomerically enriched for observed anisochrony to result. The value of $\Delta\delta$ decreases with CSA enantiomeric purity and equals zero for racemic CSA. This is a factor of considerable practical importance.

Thus far, no mention has been made of the effect of the relative proportions of the enantiomers of solute on the NMR spectra. In fact, the proportions typically have no influence whatever on the magnitude of the observed anisochrony $\Delta\delta$. However, they are directly reflected in the relative *intensities* (i.e., integrated areas) of the two observed resonances, assuming that certain experimental precautions have been taken (see Section IV). Hence, if CSA induces anisochrony, racemic solute exhibits two signals of equal intensity, enantiomerically enriched solute exhibits two signals of unequal intensity, and enantiomerically pure solute exhibits only one signal. A practical consequence of these facts is that direct enantiomeric purity determinations are possible and, in many cases, feasible.

In addition, CSA-induced anisochrony can aid in the study of intramolecular processes. When solute enantiomers are interconverted by processes such as restricted bond rotation or pyramidal inversion, the enantiomerization kinetics can be determined in favorable cases by dynamic NMR lineshape studies of solute in the presence of CSA (Küspert and Mannschreck, 1982, and references therein).

An assumption in the discussion to this point has been that solute–solute association is relatively unimportant and so can be neglected. This is probably generally appropriate for the above-discussed CSA experiments since solute–CSA complexes are typically more stable and CSA is typically used in excess (Section IV,A). In the absence of CSA, however, enantiomerically enriched mixtures of chiral solutes that are prone to associate may exhibit *self-induced anisochrony* (Williams *et al.*, 1969). Equations (6)–(8) show possible binary association equilibria for a mixture of S_d and S_l. If S_d is in large excess and $K_1 \sim$

$$S_d + S_d \overset{K_1}{\rightleftharpoons} S_dS_d \tag{6}$$

$$S_d + S_l \overset{K_2}{\rightleftharpoons} S_dS_l \tag{7}$$

$$S_l + S_l \overset{K_3}{\rightleftharpoons} S_lS_l \tag{8}$$

$K_2 \sim K_3$, then S_d will be associated mainly according to Eq. (6) and S_l will be associated mainly according to Eq. (7). Clearly, the time-average (fast-exchange) environments of S_d and S_l are different under these circumstances, and anisochrony of corresponding sensor nuclei in S_d and S_l may result. Solute S_d can be thought of as the CSA for a mixture of S_d and S_l.

III. Chiral Solvating Agents, Solutes, and Association Interactions

A fairly wide variety of combinations of solute and CSA functional group types have resulted in CSA-induced solute anisochrony. Primary literature reports (through mid-1981) of such cases are documented in Table I. Accompanying structures are shown in Fig. 1. In Table I the solutes are classified and organized

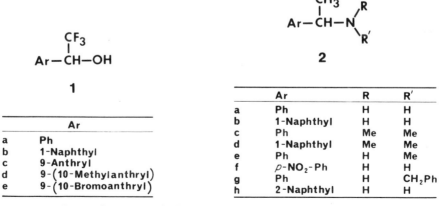

$$\underset{\textbf{1}}{\overset{\overset{\text{CF}_3}{|}}{Ar-CH-OH}}$$

Ar
a Ph
b 1-Naphthyl
c 9-Anthryl
d 9-(10-Methylanthryl)
e 9-(10-Bromoanthryl)

$$\underset{\textbf{2}}{\overset{\overset{\text{CH}_3}{|}}{Ar-CH-N{\overset{\diagup R}{\diagdown R'}}}}$$

	Ar	R	R'
a	Ph	H	H
b	1-Naphthyl	H	H
c	Ph	Me	Me
d	1-Naphthyl	Me	Me
e	Ph	H	Me
f	p-NO$_2$-Ph	H	H
g	Ph	H	CH$_2$Ph
h	2-Naphthyl	H	H

Fig. 1. Structures of CSAs and solutes. (*continued on pp. 157–159*)

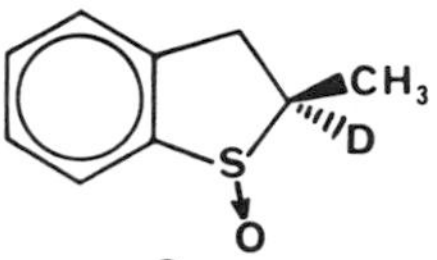

3

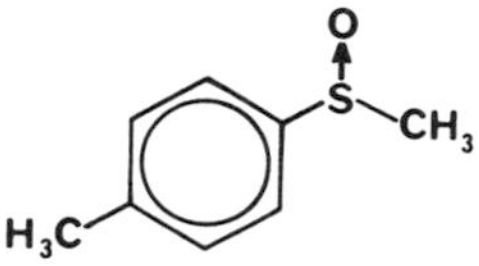

4

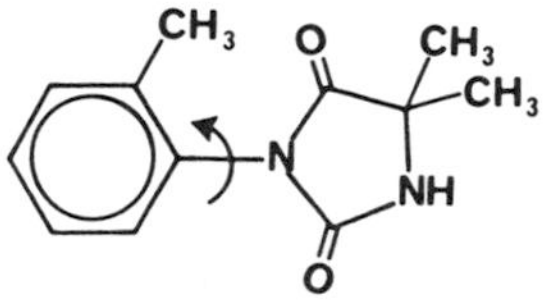

5

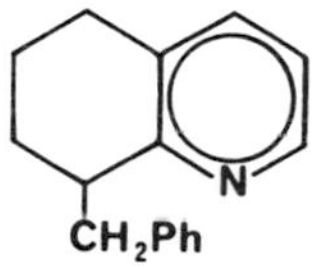

6

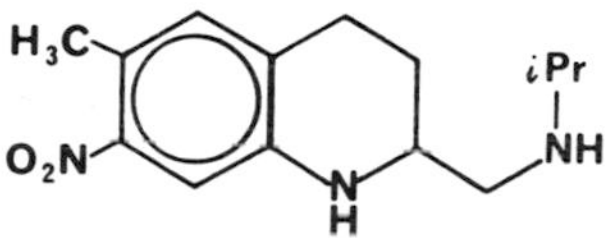

7

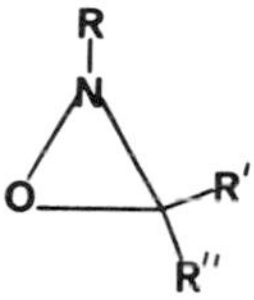

8 R = alkyl;
R', R'' = H, alkyl, Ph

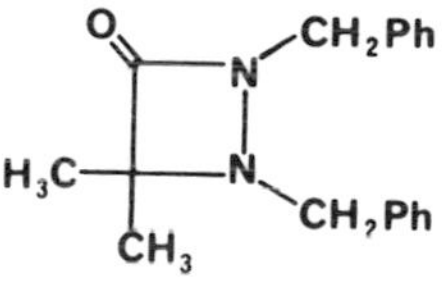

9

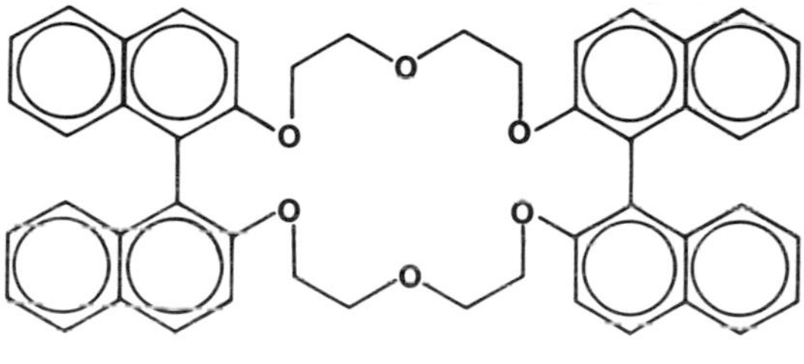

10

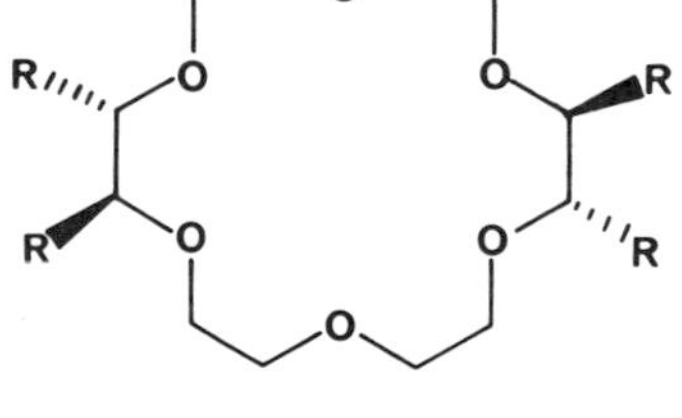

11 R = CH₂OCPh₃

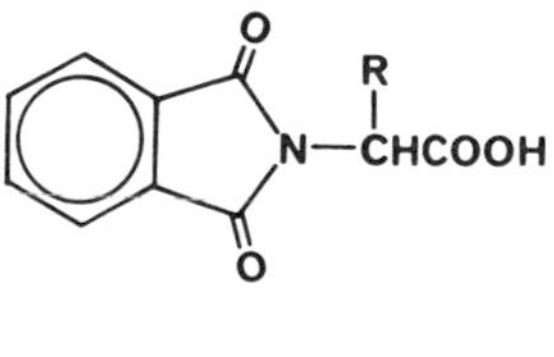

12

Fig. 1 (*Continued*)

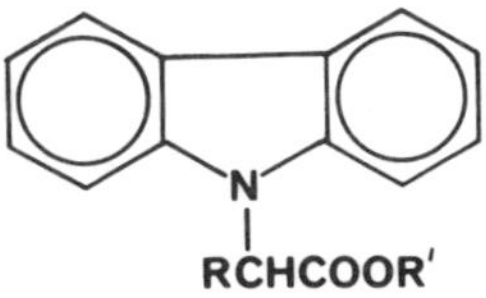

13

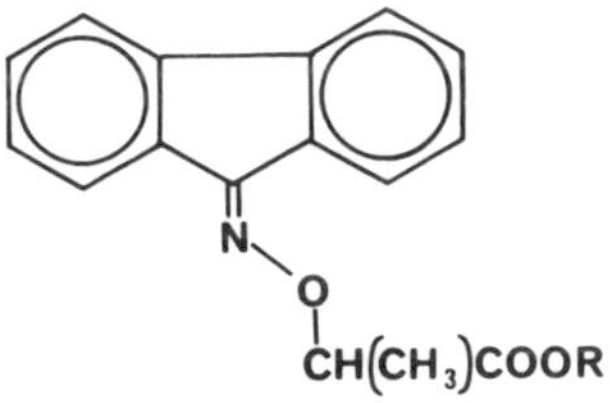

14

a R = H
b R = Me

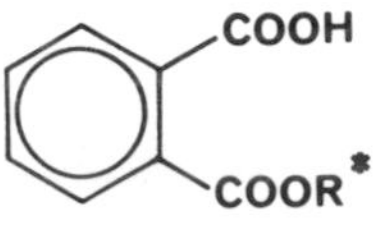

15

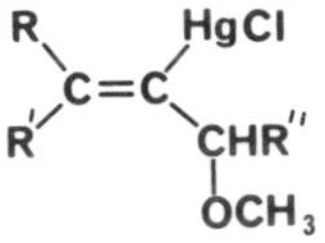

16

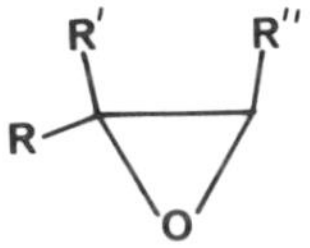

17

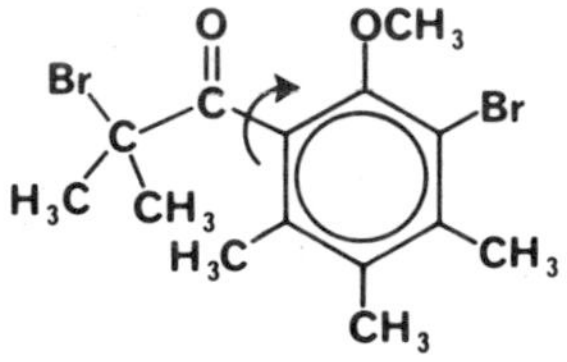

18

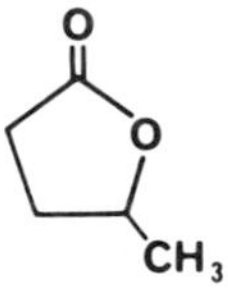

19

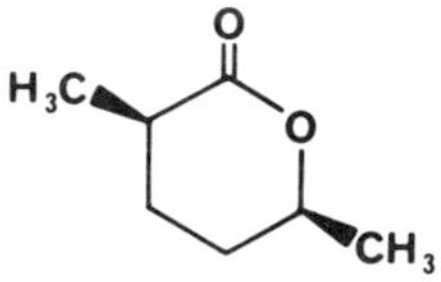

20

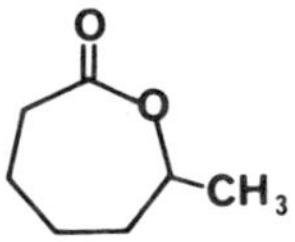

21

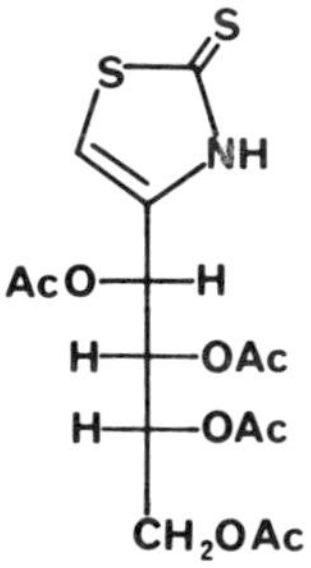

22

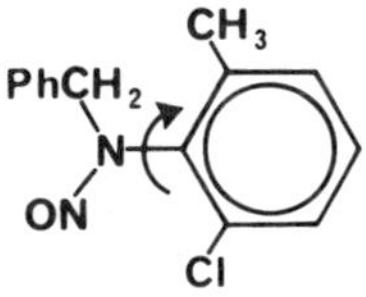

23

Fig. 1 (*Continued*)

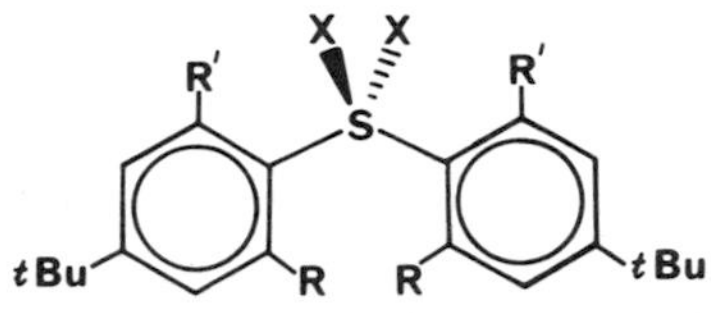

24 R=COOEt;
R′=C(CH₃)₂OH

a X = lone pair
b X=O

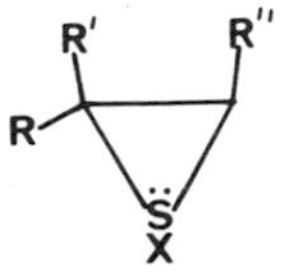

25

a X= lone pair
b X=O

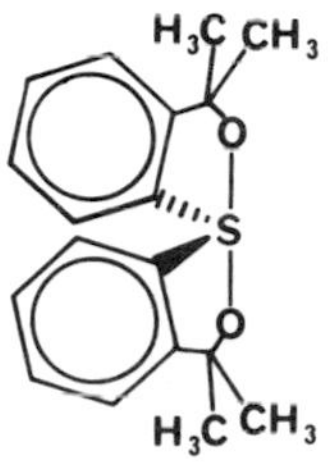

26

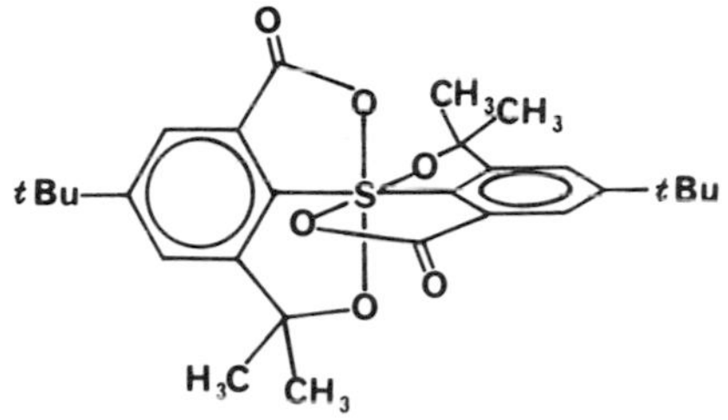

27

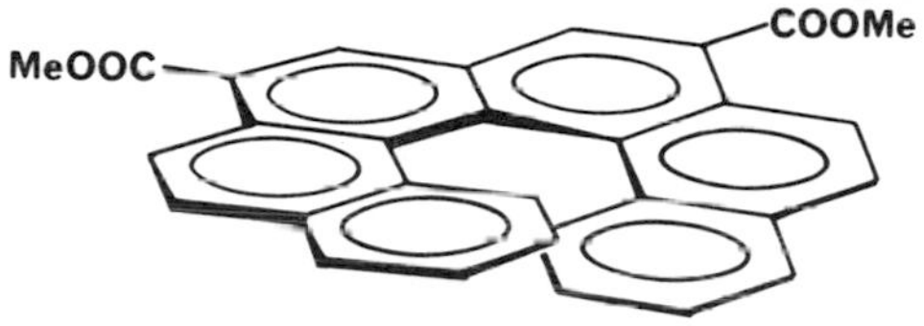

28

Fig. 1   (Continued)

according to functional group type before the addition of CSA. In some cases the assignment of a multifunctional solute to a compound class was necessarily arbitrary.

Inspection of Table I and the structures in Fig. 1 reveals that aryltrifluoromethylcarbinols (**1**) and 1-arylethylamines (**2**) have been among the most extensively employed CSAs. This derives in part from their general applicability and in part from their commercial availability in enantiomerically pure or enriched forms.[1] Both CSAs **1** and **2** rely on hydrogen bonding as the primary

[1]Enantiomers of **1a** are available from Alfa Products, Thiokol/Ventron Division, Danvers, Massachusetts, and George Uhe Co., Inc., New York, New York. Enantiomers of **1c** are available from Aldrich Chemical Co., Inc., Milwaukee, Wisconsin, and Regis Chemical, Morton Grove, Illinois. Enantiomers of **2a** and **2b** are available from a variety of suppliers.

TABLE I

Literature Reports of CSA-Induced Anisochrony in Solute NMR Spectra

Entry	Solute	CSA	Nucleus	Reference
	Alcohols			
1	Alkylarylcarbinols, $ArCH(OH)R$	**2b**	^1H	Burlingame and Pirkle (1966); Pirkle and Beare (1967)
2		**2a, 2c, 2d, 2e**	^1H, ^{13}C	Pirkle and Hoekstra (1975)
3		**3, 4**	^1H	Anet et al. (1968)
4	Arylperfluoroalkylcarbinols (e.g., **1a**)	**2a**	^{19}F	Pirkle (1966)
5		α-Cyclodextrin	^{19}F	MacNicol and Rycroft (1977)
6		**2a, 2c, 2d, 2e**	^1H, ^{13}C	Pirkle and Hoekstra (1975)
7		**2b**	^1H	Burlingame and Pirkle (1966)
8		**2b**	^{19}F	Pirkle and Beare (1967); Pirkle and Burlingame (1967)
9	Alkylperfluoroalkylcarbinols, $RCH(OH)CF_3$ (e.g., R = cyclohexyl)	**2b**	^{19}F	Pirkle and Burlingame (1967)
10	Arylhaloalkylcarbinols, $ArCH(OH)R$ (e.g., R = CH_2I, CCl_3)	**2a, 2c, 2d, 2e**	^1H, ^{13}C	Pirkle and Hoekstra (1975)
	Amides			
11	Propionamides and propionanilides	**2a**	^1F	Snatzke et al. (1973)
12	N-Alkyltrifluoroacetamides, $PhCH(CH_3)NHC(O)CF_3$	$[i\text{-}PrOC(=O)CH(R)NH]_2C=O$	^1H, ^{13}C	Lochmüller et al. (1972)
13	1- and 3-Arylhydantoins (e.g., **5**)	**1a**[a]	^1H	Colebrook et al. (1975)
	Amines			
14	2-Arylalkylamines, $ArCH(NH_2)R$	**1a, 1b**	^1H	Pirkle et al. (1968)
15		$PhCH(CH_3)COOH$[b]	^1H	Guetté et al. (1968)
16		Lactic acid, mandelic acid,[b] β-cyclodextrin hydrate	^{15}N	Dyllick-Brenzinger and Roberts (1980)
17	Tetrahydroquinoline (**6**)	$PhC(OCH_3)(CF_3)COOH$[b]	^1H	Baxter and Richard (1972)
18	Tetrahydroquinoline (**7**)	**1c, 1d, 1e**	^1H	Pirkle and Rinaldi (1977); Pirkle and Rinaldi (1978)
19	Oxaziridines (**8**)	**1a, 1b**	^1H	Forni et al. (1978)
20	Diazetidinone (**9**)	Dibenzoyltartaric acid[b]	^1H	Mannschreck et al. (1973b)

160

No.	Substrate	Chiral solvating agent	Nucleus	Reference
	Alkylammonium salts			
21	PhCH(CH₃)NH₃⁺ X⁻; X = PF₆, ClO₄, or SCN	Bisdinaphthyl 22-crown-6 ethers (e.g., **10**)	^1H	Kyba et al. (1973, 1978)
22		Carbohydrate-derived 18-crown-6 ethers (e.g., **11**)	^1H, ^{13}C	Curtis et al. (1975, 1977); Haslegrave and Stoddart (1979)
23		Carbohydrate-derived dinaphthyl 20-crown-6 ether	^1H, ^{13}C	Curtis et al. (1976)
24		Carbohydrate-derived 30-crown-10 ether	^1H	Metcalfe et al. (1981)
25		Chiral diaza crown ethers	^1H	Pearson et al. (1979); Chadwick et al. (1981)
	Amine oxides			
26	N,N-Dialkylarylamine oxides, ArN(O)RR'	**1a, 1b**	^1H	Pirkle et al. (1971)
27		**1b**	^1H	Moriwaki et al. (1976); Pirkle et al. (1969)
	Amino acid derivatives and salts			
28	α-Amino acid salts, RCH(COOH)NH₃⁺ ClO₄⁻	Bisdinaphthyl 22-crown-6 ethers (e.g., **10**)	^1H	Peacock et al. (1980)
29	α-Amino acid esters, RCH(NH₂)CO₂CH₃	**1a**	^1H	Pirkle and Beare (1969)
30	β-Amino ester, CH₃CH(NH₂)CH₂CO₂CH₃	**1a**	^1H	Pirkle and Beare (1969)
31	α-Amino ester salts, RCH(CO₂CH₃)NH₃⁻ X⁻; X = PF₆, ClO₄	Biscinaphthyl 22-crown-6 ethers (e.g., **10**)	^1H	Kyba et al. (1978); Helgeson et al. (1974); Peacock et al. (1978)
32	Pro-OCH₂Ph · HCl	Dinaphthyl 20-crown-6 esters	^1H	Lingenfelter et al. (1981)
33	Phe-OCH₃ · HCl, Val-OCH₃ · HCl	Cyclo(L-Pro-Gly)ₙ; n = 3, 4, 5	^{13}C	Madison et al. (1977); Deber and Blout (1974)
34	N-Acetyl-Phe	Chymotrypsin	^1H, ^{19}F	Spotswood et al. (1967)
35	N-Phthalimido-α-amino acids (**12**)	**2a**[b]	^1H	Ejchart et al. (1971)
36	α-Amino acid carbazole derivatives (**13**)	**14a**	^1H	Mannschreck et al. (1978)

(continued)

161

TABLE I (Continued)

Entry	Solute	CSA	Nucleus	Reference
	Carboxylic acids			
37	ArCH(R)COOH	2a, 2b[b]	¹H	Guetté et al. (1968); Horeau and Guetté (1968)
38	Phthalic acid half-esters of chiral alcohols (15)	2b[b]	¹H	Guetté et al. (1968)
39	RR'C=NOCH₂COOH (chiral ketone derivative)	2a, 2b[b]	¹H	Mamlok et al. (1971)
40	PhCH(OAc)COOH	2a[b]	¹H	Mikołajczyk et al. (1971); Ejchart and Jurczak (1971)
41	HO₂CCH₂C(CH₃)(R)COOH	SIA[c]	¹H	Horeau and Guetté (1974)
	Ethers			
42	Chloromercuriallyl methyl ethers (16) (chiral allene derivatives)	1c	¹H	Pirkle and Boeder (1977)
43	Oxiranes (17)	1a	¹H	Moretti et al. (1973)
44	Chiral 18-crown-6 ether	2a · HBr	¹H	Merz et al. (1981)
45	Chiral diaza crown ethers	2a · HSCN	¹H	Pearson et al. (1979)
	Haloalkanes			
46	(dl)-CHFClCHFCl	L-Bornyl acetate	¹⁹F	Abraham et al. (1974); Abraham and Siverns (1973)
	Hydroxy esters			
47	RC(Ph)(OH)COOCH₃	2b	¹H, ¹⁹F	Pirkle and Beare (1968b)
	Ketones			
48	Camphor	1a	¹H	Jochims et al. (1967)
49	Substituted acetophenone (18)	1a[a]	¹H	Küspert and Mannschreck (1982)
	Lactones			
50	γ-Lactones (e.g., 19)	1c	¹H, ¹³C	Pirkle et al. (1977); Pirkle and Sikkenga (1977); Epstein and Gaudioso (1979); Pirkle and Adams (1980a)

51	δ-Lactones (e.g., **20**)	1c	^{1}H	Pirkle and Adams (1978, 1979, 1980b)
52	ε-Lactones (e.g., **21**)	1c	^{1}H	Pirkle and Adams (1979)
	Natural products and derivatives			
53	Cocaine	1a	^{1}H	Jochims et al. (1967)
54	Dihydroquinine	SIA[c]	^{1}H	Williams et al. (1969)
55	Glucosamine derivative (**22**)	1a	^{1}H	Jochims et al. (1967)
	Nitrosamines			
56	Alkylarylnitrosamine (**23**)	1a[a]	^{1}H	Mannschreck et al. (1973a,b)
	Organophosphorus compounds			
57	Phosphine oxides, $PhP(O)(CH_3)R$	1a, 1b	^{1}H	Pirkle et al. (1969)
58	Alkylphenylphosphinothioic acids, $RPhP(S)OH$	2a, 2b[b]	^{1}H, ^{13}C, ^{31}P	Mikołajczyk et al. (1970, 1978)
59		SIA[c] or another $RPhP(S)OH$	^{1}H	Harger (1978a)
60	Alkyl alkylphenylphosphinates, $RPhP(O)OR'$	$RPhP(S)OH$; R = Me, tBu	^{1}H	Harger (1978b)
61	S-Alkyl alkylphenylphosphinothioates, $RPhP(O)SR'$	$RPhP(S)OH$; R = Me, tBu	^{1}H	Harger (1978b)
62	Alkylphenylphosphinamides, $RPhP(O)NHR'$; R' = H, alkyl	$RPhP(S)OH$; R = Me, tBu	^{1}H	Harger (1978a)
63		SIA[c] or another $RPhP(O)NHR$[l]	^{1}H	Harger (1976, 1977)
64	Alkyl hydrogen alkylphosphonothioates, $R(R'O)P(S)OH$	2a–2c, 2f–2h, $PhHC=NCH(Ph)CH_3$[b]	^{1}H, ^{13}C, ^{31}P	Mikołajczyk et al. (1970, 1978); Mikołajczyk and Omelánczuk (1972)
65	Alkyl hydrogen phenylphosphonothioates, $Ph(RO)P(S)OH$	2a[b]	^{1}H	Mikołajczyk et al. (1978)
66	Alkylaryl hydrogen phosphorothioates, $(RO)(ArO)P(O)OH$	$RPhP(S)OH$; R = Me, tBu	^{1}H	Harger (1978a)
67		2a[b]	^{1}H, ^{31}P	Mikołajczyk et al. (1978)
68	S,O-Dialkyl hydrogen phosphorothiolothionates, $(RO)(RS)P(S)OH$	2a[b]	^{1}H	Mikołajczyk et al. (1978)
69	Pyrophosphoramides, $R(NMe_2)P(O)OP(O)(NMe_2)F$	1a	^{31}P	Joesten et al. (1973)

(continued)

TABLE I (*Continued*)

Entry	Solute	CSA	Nucleus	Reference
	Organosulfur compounds			
70	Diaryl sulfides (e.g., **24a**)	**1a**[d]	^{1}H	Lam and Martin (1981a)
71	Episulfides (**25a**)	**1a**	^{1}H	Bucciarelli *et al.* (1977)
72	Dialkyl sulfoxides, RS(O)R'	**1a, 1c**	^{1}H	Pirkle *et al.* (1974); Pirkle and Pavlin (1974); Pirkle and Beare (1968a)
73	Alkylaryl sulfoxides, ArS(O)R	**1a**	^{1}H	Pirkle and Beare (1968a); Whitney and Pirkle (1974)
74	Episulfoxides (**25b**)	**1a**	^{1}H	Bucciarelli *et al.* (1977)
75	Sulfinate esters (acyclic and cyclic), RS(O)OR', ArS(O)OR	**1a, 1b, 1d, 1e,** PhCH(CF$_2$CF$_2$CF$_3$)OH	^{1}H	Pirkle and Hoekstra (1976); Pirkle *et al.* (1969)
76	Thiosulfinates, RS(O)SR'	**1a**	^{1}H	Pirkle *et al.* (1969)
77	Sulfinamides, RS(O)NR$_2'$	**1a**	^{1}H	Pirkle *et al.* (1969)
78	Sulfites, ROS(O)OR	**1a**	^{1}H	Pirkle *et al.* (1969)
79	Diarylsulfone (**24b**)	**1a**[d]	^{1}H	Lam and Martin (1981a)
80	Spirosulfuranes (e.g., **26**)	**1a**[d]	^{1}H	Lam and Martin (1981b); Martin and Adzima (1977); Adzima *et al.* (1978)
81	Persulfurane (**27**)	**1a**[d]	^{1}H	Lam and Martin (1977, 1980)
	Miscellaneous			
82	π-Acid (**14b**)	Hexahelicene (**28**)	^{1}H	Balan and Gottlieb (1981)
83	Organothiophosphorus depsipepsides, (EtO)P(=Y)(Me)SCH$_2$C(=O)NRCH(*i*Pr)C(=O)X; Y = S, O; X = OH, OEt; R = H, Me	SIA[c]	^{31}P	Kabachnik *et al.* (1978)

[a] Kinetic study of solute atropisomerism.
[b] Salt formation upon reaction with solute.
[c] Self-induced anisochrony of enantiomerically enriched solute.
[d] Chirality demonstration on racemic solute.

solute-binding force. Generally, if the solute is a hydrogen bond acceptor (e.g., a tertiary amine), the CSA of choice is an efficient hydrogen bond donor (1), and if the solute is a hydrogen bond donor (e.g., an alcohol), the CSA of choice is a hydrogen bond acceptor (2). Ultimately, the applicability of CSA types 1 and 2 derives not only from their solvation capabilities but also from the fact that they possess aryl groups directly attached to their chiral centers. Differences in the local magnetic environments of the sensor nuclei in diastereomeric association complexes result at least in part from the different average proximities and orientations of the nuclei relative to these magnetically anisotropic aryl groups. This is particularly important when ^{1}H-NMR spectroscopy is employed. Thus 1b often induces a larger $\Delta\delta$ in the ^{1}H-NMR spectrum of an appropriate solute than does 1a (Pirkle and Hoekstra, 1976), and 1c has been promoted as the best of these three aryltrifluoromethylcarbinols (Pirkle and Hoover, 1982). Alkyltrifluoromethylcarbinols have generally been found to be less useful than 1 for the induction of anisochrony (Pirkle et al., 1971). Solvates involving 1 or 2 may, according to the solute, be conformationally biased by secondary interactions between solute and CSA, thereby influencing the $\Delta\delta$ magnitudes. For the case of association complex formation between 1 and a solute possessing two basic sites, Pirkle has proposed a solvation model in which the secondary interaction is hydrogen bonding between the carbinyl hydrogen of 1 and the least basic of the two solute basic sites. This model has proved useful in predicting absolute configurations of dibasic solutes (Pirkle and Hoover, 1982).

Several types of interactions other than hydrogen bonding may contribute primarily or secondarily to CSA–solute association, and these are illustrated by examples from Table I. Anisochrony has been induced through the formation of diastereomeric charge-transfer (π-acid–π-base) complexes (entries 36 and 82). The efficient binding of chiral alkylammonium ions (including amino acid conjugate acids and derivatives) by various types of chiral crown ethers (entries 21, 25, 28, 31, 32, 44, and 45) is due not only to hydrogen bonding but also to pole–dipole attractions as well as charge-transfer interactions in some cases (Cram, 1976; Cram and Cram, 1978; Stoddart, 1979, 1981). Van der Waals attractions and release of high-energy water molecules each contribute to the inclusion of lipophilic solute guests by cyclodextrins in aqueous media (entries 5 and 16) (Bender and Komiyama, 1978). Finally, ion pairing may be a principle contributor to binding in certain cases (entries 15–17, 20, 35, 37–40, 58, 64, 65, 67, and 68). In such cases enantiomerically pure CSA and solute enantiomers react by proton transfer to form diastereomeric salts. The salts may exhibit different NMR spectra in solution under conditions that promote ion pairing between solute conjugate acid (or base) and CSA conjugate base (or acid) counterion.

IV. Practical Experimental Considerations

A. General

The experimental method for CSA–NMR determination of enantiomeric purity is straightforward. A solution consisting of solute, CSA, and usually an achiral cosolvent or diluent is prepared and the NMR spectrum is recorded. Racemic solute or a mixture of known enantiomeric purity is examined first, because an observable $\Delta\delta$ must be demonstrated. Several sample variables other than the identities of solute and CSA affect the magnitude of observed $\Delta\delta$ values. Generally, conditions that maximize association complex formation are most useful for maximizing $\Delta\delta$. Thus CSA concentration is usually chosen to be three to five times that of solute. The achiral cosolvents (or diluents, if the CSA is a solid) used in conjunction with the most common CSAs (1 and 2) are aprotic and relatively nonpolar if possible so that competitive association with either solute or CSA is avoided. $CDCl_3$, CD_2Cl_2, CCl_4, CS_2, and [$^2H_6$]benzene (i.e., benzene-d_6) have all been used successfully. [$^2H_6$]Benzene has been recommended as the diluent of choice for 1c (Pirkle and Hoover, 1982), and pyridine has been found useful for the salts formed by reaction of phosphorus acids with 2 (Mikołajczyk et al., 1978). Lower sample temperature usually favors association, and several examples of $\Delta\delta$ increases with lower temperatures have been reported (Pirkle and Burlingame, 1967; Pirkle and Hoekstra, 1975).

B. Proton Nuclear Magnetic Resonance

The advantages of using ^{1}H-NMR spectroscopy for measuring enantiomeric purity lie in the ubiquity and high sensitivity of the ^{1}H nucleus and in the fact that relative signal intensities directly reflect relative numbers of resonating nuclei and hence relative enantiomeric populations.

The reliability of the CSA–^{1}H-NMR method is dependent on the appearance and quality of spectra and on the treatment of the spectral data. Quantitative analysis by NMR has been reviewed (Kasler, 1973; Leyden and Cox, 1977; Martin et al., 1980, p. 350).

Generally, a high signal-to-noise ratio (S/N) and baseline resolution of the two signals to be compared are desirable. Adequate S/N is generally not a problem in ^{1}H-NMR analysis, even for small samples. Although ^{1}H $\Delta\delta$ values of $\sim$0.01–0.15 ppm have been reported $\Delta\nu$ values of less than 4 Hz (at 100 MHz) are most typical. Therefore, the spectrometer magnetic field homogeneity should be carefully tuned for maximum resolution. Analysis is most simply and reliably achieved by intensity comparison of singlets, although multiplets may be compared in cases in which $\Delta\nu$ magnitude permits this (Snatzke et al., 1973).

Observed $\Delta\nu$ values (hertz) may, of course, be increased by the use of a higher-field spectrometer. Precautions should be taken to ensure the recording of accurate lineshapes. If the spectrum is obtained in the continuous wave (CW) mode, a narrow frequency range should be observed, and appropriate sweep rate and RF power should be used to avoid saturation effects. If the spectrum is obtained in the Fourier transform (FT) mode, the digital resolution must be high enough that each peak is accurately defined by a sufficient number of data points (Martin *et al.*, 1980, p. 364).

For the actual measurement of the relative intensities, several methods are available. Cutting and accurate weighing of the peak areas has been reported to be the method of choice for CW ^{1}H-NMR studies (Martin *et al.*, 1980, p. 368). Electronic integration yields area ratios of resolved signals with a maximum accuracy of about $\pm 1\%$ (for a 1 : 1 ratio). Pirkle and co-workers have found that relative peak heights differ little from relative peak areas and have encouraged the use of the former for convenience (Pirkle and Beare, 1969, Pirkle and Hoover, 1982). Others, however, have cautioned against the use of peak height ratios in ^{1}H-NMR analysis because of possible linewidth variation between the resonances (Martin *et al.*, 1980, p. 368). Irrespective of the method used, multiple measurements should be made, and the precision and accuracy should be checked by measurements on a sample of known enantiomeric purity. On the basis of such checks that have been published (Pirkle and Beare, 1969, Pirkle and Rinaldi, 1977), precision of $\pm 1\%$ and accuracy of $\pm$ 1 to 3% can generally be expected for percent enantiomeric excess (% ee) values carefully measured by this technique. For very high or very low % ee, accuracy is no doubt lower.

C. Carbon-13 and Heteroatom Nuclear Magnetic Resonance

The relatively few reported cases of CSA-induced anisochrony in ^{13}C-NMR or heteronuclear spectra are listed in Table I. Ranges of $\Delta\delta$ values that have been reported are 0.04–2.47 ppm for ^{13}C (entries 2, 10, 12, 22, 23, 33, 50, 58, and 64), 0.16–0.68 ppm for ^{15}N (entry 16), 0.018–0.062 ppm for ^{19}F (entries 4, 5, 8, 9, 46, and 47), and 0.025–1.00 ppm for ^{31}P (entries 58, 64, 67, 69, and 83). Carbon-13 and heteroatom NMR spectra are typically obtained in the FT mode. The low sensitivities of ^{31}P and especially ^{13}C and ^{15}N relative to ^{1}H make multiple-pulse experiments a necessity if sufficient S/N is to be obtained. Broadband ^{1}H decoupling is normally employed for ^{13}C, ^{15}N, and ^{31}P. Thus, all of the signals to be compared are singlets, a real advantage in enantiomeric purity measurements (see preceding discussion).

For ^{13}C, ^{15}N, and ^{31}P nuclei, the recorded signal intensities from a typical steady-state multiple-pulse FT experiment depend not only on numbers of resonating nuclei, but also on their longitudinal relaxation times (T_1) and, if broadband decoupling is employed, on nuclear Overhauser effects (NOE). When the signals to be compared arise from very similar nuclei, however, as in CSA–NMR experiments, T_1 and NOE differences are typically minimal (Heimstra and Wynberg, 1977, and references therein). In any case, T_1 effects and NOE can be eliminated if necessary by the use of long delays ($5T_1$) between pulses and gated decoupling, respectively (Martin *et al.*, 1980). The small $\Delta\delta$ values encountered in the CSA–NMR experiments should generally preclude intensity variations due to nonuniform pulse power (Martin *et al.*, 1980, p. 108).

Since linewidths are determined principally by ^{1}H decoupling and by the S/N-enhancing mathematical weighting functions usually applied to the free induction decay before Fourier transformation, relative peak heights may be taken as reliable indicators of relative areas and hence of enantiomeric populations if care is taken to ensure adequate digital resolution.

V. Comparison with the Chiral Lanthanide Shift Reagent Method

The use of chiral lanthanide shift reagents (CLSRs) for NMR determination of enantiomeric purity has been reviewed by Fraser (Chapter 9, this volume). Many of the principles of the method are similar to those of the CSA method. Typically, CLSRs induce larger chemical shift differences than do CSAs and thus have received more use for % ee determinations, despite the fact that they induce nonuniform line broadening of resonances. [Several reports of the use of CSAs in combination with achiral lanthanide shift reagents have demonstrated the expected increases in anisochrony magnitudes relative to the use of CSAs alone (Jennison and Mackay, 1973; McCreary *et al.*, 1974; Pirkle and Sikkenga, 1975; Ajisaka and Kainosho, 1975).] The CLSRs are less useful than CSAs for NMR determinations of absolute configurations (Pirkle and Hoover, 1982).

VI. Conclusions

The CSA–NMR method is a relatively general, direct spectroscopic method for the determination of enantiomeric purity. The method is inexpensive and applicable to a variety of functional group types. This author is of the opinion that the method has been underutilized to date. The commercial availability of **1c** and the increasingly routine use of high-field ^{1}H-NMR and ^{13}C-NMR spectroscopy by organic chemists may prompt greater utilization of the technique.

References

Abraham, R. J., and Siverns, T. M. (1973). *Org. Magn. Reson.* **5,** 253.

Abraham, R. J., Cooper, M. A., Siverns, T. M., Swinton, P. F., Weder, H. G., and Cavalli, L. (1974). *Org. Magn. Reson.* **6,** 331.

Adzima, L. J., Chiang, C. C., Paul, I. C., and Martin, J. C. (1978). *J. Am. Chem. Soc.* **100,** 953.

Ajisaka, K., and Kainosho, M. (1975). *J. Am. Chem. Soc.* **97,** 1761.

Anet, F. A. L., Sweeting, L. M., Whitney, T. A., and Cram, D. J. (1968). *Tetrahedron Lett.,* p. 2617.

Balan, A., and Gottlieb, H. E. (1981). *J. Chem. Soc. Perkin Trans.* 2, p. 350.

Baxter, C. A. R., and Richard, H. C. (1972). *Tetrahedron Lett.,* p. 3357.

Bender, M. L., and Komiyama, M. (1978). "Cyclodextrin Chemistry." Springer–Verlag, Berlin.

Bucciarelli, M., Forni, A., Moretti, I., and Torre, G. (1977). *Tetrahedron* **33,** 999.

Burlingame, T. G., and Pirkle, W. H. (1966). *J. Am. Chem. Soc.* **88,** 4294.

Chadwick, D. J., Cliffe, I. A., and Sutherland, I. O. (1981). *J. Chem. Soc. Chem. Commun.,* p. 992.

Colebrook, L. D., Icli, S., and Hund, F. H. (1975). *Can. J. Chem.* **53,** 1556.

Cram, D. J. (1976). *In* "Applications of Biochemical Systems in Organic Chemistry" (J. B. Jones, C. J. Sih, and D. Perlman, eds.), Part 2, p. 815. Wiley (Interscience), New York.

Cram, D. J., and Cram, J. M. (1978). *Acc. Chem. Res.* **11,** 8.

Curtis, W. D., Laidler, D. A., Stoddart, J. F., and Jones, G. H. (1975). *J. Chem. Soc. Chem. Commun.,* p. 835.

Curtis, W. D., King, R. M., Stoddart, J. F., and Jones, G. H. (1976). *J. Chem. Soc. Chem. Commun.,* p. 284.

Curtis, W. D., Laidler, D. A., Stoddart, J. F., and Jones, G. H. (1977). *J. Chem. Soc. Perkin Trans. 1,* p. 1756.

Deber, C. M., and Blout, E. R. (1974). *J. Am. Chem. Soc.* **96,** 7566.

Dyllick-Brenzinger, R., and Roberts, J. D. (1980). *J. Am. Chem. Soc.* **102,** 1166.

Ejchart, A., and Jurczak, J. (1970). *Wiad. Chem.* **24,** 857.

Ejchart, A., and Jurczak, J. (1971). *Bull. Acad. Pol. Sci Ser. Sci. Chim.* **19,** 725.

Ejchart, A., Jurczak, J., and Bankowski, K. (1971). *Bull. Acad. Pol. Sci. Ser. Sci. Chim.,* **19,** 731.

Epstein, W. W., and Gaudioso, L. A. (1979). *J. Org. Chem.* **44,** 3113.

Erdik, E. (1981). *Doga Ser. B* **5,** 53.

Forni, A., Moretti, I., and Torre, G. (1978). *Tetrahedron Lett.,* p. 2941

Gaudemer, A. (1977). *In* "Stereochemistry, Fundamentals and Methods" (H. B. Kagan, ed.), Vol. 1, p. 44. Thieme, Stuttgart.

Guetté, J.-P., Lacombe, L., and Horeau, A. (1968). *C. R. Hebd. Seances Acad. Sci. Ser. C* **267,** 166.

Harger, M. J. P. (1976). *J. Chem. Soc. Chem. Commun.,* p. 555.

Harger, M. J. P. (1977). *J. Chem. Soc. Perkin Trans.* 2, p. 1882.

Harger, M. J. P. (1978a). *J. Chem. Soc. Perkin Trans.* 2, p. 326.

Harger, M. J. P. (1978b). *Tetrahedron Lett.,* p. 2927.

Haslegrave, J. A., and Stoddart, J. F. (1979). *Tetrahedron Lett.,* p. 2279.

Heimstra, H., and Wynberg, H. (1977). *Tetrahedron Lett.,* p. 2183.

Helgeson, R. C., Timko, J. M., Moreau, P., Peacock, S. C., Mayer, J. M., and Cram, D. J. (1974). *J. Am. Chem. Soc.* **96,** 6762.

Horeau, A., and Guetté, J.-P. (1968). *C. R. Hebd. Seances Acad. Sci. Ser. C* **267,** 257.

Horeau, A., and Guetté, J.-P. (1974). *Tetrahedron* **30,** 1923.

Jacques, J., Collet, A., and Wilen, S. H. (1981). "Enantiomers, Racemates, and Resolutions." Wiley, New York.

Jennison, C. P. R., and Mackay, D. (1973). *Can. J. Chem.* **51**, 3726.

Jochims, J. C., Taigel, G., and Seeliger, A. (1967). *Tetrahedron Lett.*, p. 1901.

Joesten, M. D., Smith, H. E., and Vix, V. A. (1973). *J. Chem. Soc. Chem. Commun.*, p. 18.

Kabachnik, M. I., Mastryukova, T. A., Fedin, E. I., Vaisberg, M. S., Morozov, L. L., and Petrovskii, P. V. (1978). *Russ. Chem. Rev. (Engl. Transl.)* **47**, 821.

Kasler, F. (1973). "Quantitative Analysis by NMR Spectroscopy." Academic Press, New York.

Küspert, R., and Mannschreck, A. (1982). *Org. Magn. Reson.* **19**, 6.

Kyba, E. P., Koga, K., Sousa, L. R., Siegel, M. G., and Cram, D. J. (1973). *J. Am. Chem. Soc.* **95**, 2692.

Kyba, E. P., Timko, J. M., Kaplan, L. J., deJong, F., Gokel, G. W., and Cram, D. J. (1978). *J. Am. Chem. Soc.* **100**, 4555.

Lam, W. Y., and Martin, J. C. (1977). *J. Am. Chem. Soc.* **99**, 1659.

Lam, W. Y., and Martin, J. C. (1980). *J. Am. Chem. Soc.* **103**, 120.

Lam, W. Y., and Martin, J. C. (1981a). *J. Org. Chem.* **46**, 4458.

Lam, W. Y., and Martin, J. C. (1981b). *J. Org. Chem.* **46**, 4468.

Leyden, D. E., and Cox, R. H., (1977) "Analytical Applications of NMR." Wiley (Interscience), New York.

Lingenfelter, D. S., Helgeson, R. C., and Cram, D. J. (1981). *J. Org. Chem.* **46**, 393.

Lochmüller, C. H., Harris, J. M., and Souter, R. W. (1972). *J. Chromatogr.* **71**, 405.

MacNicol, D. D., and Rycroft, D. S. (1977). *Tetrahedron Lett.*, p. 2173.

Madison, V., Deber, C. M., and Blout, E. R. (1977). *J. Am. Chem. Soc.* **99**, 4788.

Mamlok, L., Marquet, A., and Lacombe, L. (1971). *Tetrahedron Lett.*, p. 1039.

Mannschreck, A., Jonas, V., and Kolb, B. (1973a). *Angew. Chem. Int. Ed. Engl.* **12**, 909 (*Angew. Chem.* **85**, 994).

Mannschreck, A., Jonas, V., and Kolb, B. (1973b). *Angew. Chem. Int. Ed. Engl.* **12**, 583 (*Angew. Chem.* **85**, 590).

Mannschreck, A., Rosa, P., Brockmann, H., Jr., and Kenner, T. (1978). *Angew. Chem. Int. Ed. Engl.* **17**, 940 (*Angew. Chem.* **90**, 995).

Martin, J. C., and Adzima, L. J. (1977). *J. Org. Chem.* **42**, 4006.

Martin, M. L., Delpuech, J.-J., and Martin, G. J. (1980). "Practical NMR Spectroscopy." Heyden, London.

McCreary, M. D., Lewis, D. W., Wernick, D. L., and Whitesides, G. M. (1974). *J. Am. Chem. Soc.* **96**, 1038.

Merz, A., Eichner, M., and Tomahough, R. (1981). *Liebigs Ann. Chem.*, p. 1774.

Metcalfe, J. C., Stoddart, J. F., Jones, G., Crawshaw, T. H., Gavuzzo, E., and Williams, D. J. (1981). *J. Chem. Soc. Chem. Commun.*, p. 432.

Mikołajczyk, M., and Omelańczuk, J. (1972). *Tetrahedron Lett.*, p. 1539.

Mikołajczyk, M., Para, M., Ejchart, A., and Jurczak, J. (1970). *J. Chem. Soc. Chem. Commun.*, p. 654.

Mikołajczyk, M., Ejchart, A., and Jurczak, J. (1971). *Bull. Acad. Pol. Sci. Ser. Sci. Chem.* **19**, 721.

Mikołajczyk, M., Omelańczuk, J., Leitloff, M., Drabowicz, J., Ejchart, A., and Jurczak, J. (1978). *J. Am. Chem. Soc.* **100**, 7003.

Mislow, K., and Raban, M. (1967). *Top. Stereochem.* **1**, 1.

Moretti, I., Taddei, F., and Torre, G. (1973). *J. Chem. Soc. Chem. Commun.*, p. 25

Moriwaki, M., Yamamoto, Y., Oda, J., and Inouye, Y. (1976). *J. Org. Chem.* **41**, 300.

Peacock, S. C., Domeier, L. A., Gaeta, F. C. A., Helgeson, R. C., Timko, J. M., and Cram, D. J. (1978). *J. Am. Chem. Soc.*, **100**, 8190.

Peacock S. C., Walba, D. M., Gaeta, F. C. A., Helgeson, R. C., and Cram, D. J. (1980). *J. Am. Chem. Soc.* **102**, 2043.

Pearson, D. P. J., Leigh, S. J., and Sutherland, I. O. (1979). *J. Chem. Soc. Perkin Trans. 1*, p. 3113.

Pirkle, W. H. (1966). *J. Am. Chem. Soc.* **88**, 1837.
Pirkle, W. H., and Adams, P. E. (1978). *J. Org. Chem.* **43**, 378.
Pirkle, W. H., and Adams, P. E. (1979). *J. Org. Chem.* **44**, 2169.
Pirkle, W. H., and Adams, P. E. (1980a). *J. Org. Chem.* **45**, 4111.
Pirkle, W. H., and Adams, P. E. (1980b). *J. Org. Chem.* **45**, 4117.
Pirkle, W. H., and Beare, S. D. (1967). *J. Am. Chem. Soc.* **89**, 5485.
Pirkle, W. H., and Beare, S. D. (1968a). *J. Am. Chem. Soc.* **90**, 6250.
Pirkle, W. H., and Beare, S. D. (1968b). *Tetrahedron Lett.*, p. 2579.
Pirkle, W. H., and Beare, S. D. (1969). *J. Am. Chem. Soc.* **91**, 5150.
Pirkle, W. H., and Boeder, C. W. (1977). *J. Org. Chem.* **42**, 3697.
Pirkle, W. H., and Burlingame, T. G. (1967). *Tetrahedron Lett.*, p. 4039.
Pirkle, W. H., and Hoekstra, M. S. (1975). *J. Magn. Reson.* **18**, 396.
Pirkle, W. H., and Hoekstra, M. S. (1976). *J. Am. Chem. Soc.* **98**, 1832.
Pirkle, W. H., and Hoover, D. J. (1982). *Top. Stereochem.* **13**, 263.
Pirkle, W. H., and Pavlin, M. S. (1974). *J. Chem. Soc. Chem. Commun.*, p. 274.
Pirkle, W. H., and Rinaldi, P. L. (1977). *J. Org. Chem.* **42**, 3217.
Pirkle, W. H., and Rinaldi, P. L. (1978). *J. Org. Chem.* **43**, 4475.
Pirkle, W. H., and Sikkenga, D. L. (1975). *J. Org. Chem.* **40**, 3430.
Pirkle, W. H., and Sikkenga, D. L. (1977). *J. Org. Chem.* **42**, 1370.
Pirkle, W. H., Burlingame, T. G., and Beare, S. D. (1968). *Tetrahedron Lett.*, p. 5849.
Pirkle, W. H., Beare, S. D., and Muntz, R. L. (1969). *J. Am. Chem. Soc.* **91**, 4575.
Pirkle, W. H., Muntz, R. L., and Paul, I. C. (1971). *J. Am. Chem. Soc.* **93**, 2817.
Pirkle, W. H., Beare, S. D., and Muntz, R. L. (1974). *Tetrahedron Lett.*, p. 2295.
Pirkle, W. H., Sikkenga, D. L., and Pavlin, M. S. (1977). *J. Org. Chem.* **42**, 384.
Raban, M., and Mislow, K. (1965). *Tetrahedron Lett.*, p. 4249
Snatzke, G., Fox, J. E., and El-Abadelah, M. M. (1973). *Org. Magn. Reson.* **5**, 413.
Spotswood, T. McL., Evans, J. M., and Richards, J. H. (1967). *J. Am. Chem. Soc.* **89**, 5052.
Stoddart, J. F. (1979). *Chem. Soc. Rev.* **8**, 85.
Stoddart, J. F. (1981). *Prog. Macrocyclic Chem.* **2**, 173.
Whitney, T. A., and Pirkle, W. H. (1974). *Tetrahedron Lett.*, p. 2299.
Williams, T., Pitcher, R. G., Bommer, P., Gutzwiller, J., and Uskokovic, M. (1969). *J. Am. Chem. Soc.* **91**, 1871.

9

Nuclear Magnetic Resonance Analysis Using Chiral Shift Reagents

Robert R. Fraser

Ottawa–Carleton Institute for Graduate Research in Chemistry
University of Ottawa
Ottawa, Ontario, Canada

I. Introduction

More than a decade has elapsed since the first report appeared (Hinckley, 1969) describing the capacity of a "shift reagent" to spread out and thereby simplify proton magnetic resonance spectra. Indeed, the subsequent surge of efforts to explore the potential of this NMR technique eventually exceeded by far that which was subsequently established as the true value of shift reagents. As often happens in any initially "hot" or competitive area of research, the excessive popularity and concomitant duplication of effort was eventually followed by a back lash effect. This has led to such a diminished position of recognition that, currently, studies and uses of shift reagents are far from routine. It is the aim of this chapter to remind organic chemists that one of the most important developments in shift reagent research, the preparation and utilization of chiral shift reagents (CSRs), constitutes a basic probe for stereochemical information (Whitesides and Lewis, 1970). The important aspects of CSRs are presented using representative examples from the literature. The coverage is not meant to be exhaustive.

II. General Features of Shift Reagents

Shift reagents have been the subject of over 1000 articles since 1969. The many reviews of the topic (Von Ammon and Fischer, 1972; Reuben, 1973; Hofer, 1980) and one monograph (Sievers, 1973) provide thorough coverage of their properties, and one article has specifically reviewed CSRs (Sullivan, 1976). Nevertheless, it is instructive to recall the fundamental characteristics of achiral shift reagents before the particular features of CSRs are described.

The reaction of a 1,3-diketone with a trivalent lanthanide halide or nitrate in a basic protic medium produces a lanthanide metal trisdiketonate, or "shift reagent," as indicated in

$$\text{LaX}_3 + 3\text{ R—COCH}_2\text{COR}' \rightarrow \text{La}[\text{R—}\overset{\text{O}}{\overset{\|}{\text{C}}}\text{—CH—}\overset{\text{O}}{\overset{\|}{\text{C}}}\text{—R}']_3 + 3\text{ HX} \qquad (1)$$

The product, itself a hexacoordinate complex, is a Lewis acid capable of forming weak addition complexes with a wide variety of organic bases. Hinckley first described the effect of complex formation between cholesterol as base and Eu(dpm)$_3$ (dpm = dipivaloylmethanato ligand) on the NMR spectrum of cholesterol (Hinckley, 1969). However, it was a subsequent report (Sanders and Williams, 1970) that aroused the interest of organic chemists by demonstrating the very substantial effect of the addition of Eu(dpm)$_3$ on the NMR spectrum of hexanol. Figure 1 shows the spectrum in the presence of Eu(dpm)$_3$, which clearly demonstrates the downfield shifts, the magnitudes of which reflect the distance of each type of proton from the donor hydroxyl group. Each downfield shift, $\Delta\delta$, increases with the addition of more Eu(dpm)$_3$, reaching a limit that is termed the *bound shift*, that is, the shift difference for a signal in the hexanol–Eu(dpm)$_3$ complex versus free hexanol. Bound shifts are usually caused by a large difference in the magnetic susceptibility tensors for the hepta-coordinate

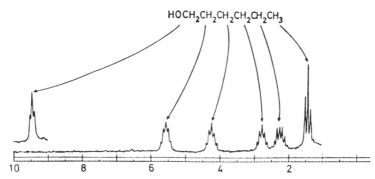

Fig. 1. 100-MHz ^{1}H-NMR spectrum of *n*-hexanol in CCl_4 after the addition of 0.29 equivalent of Eu(dpm)$_3$. The superimposed trace at left is offset by 1 ppm. From Sanders and Williams (1970).

complex. In such cases the McConnell equation provides a qualitative description: $\Delta\delta = k(1 - 3\cos^2 \theta)r^{-3}$, the shifts being inversely proportional to r^3 and proportional to an angular term, $1 - 3\cos^2 \theta$, where θ is the number of degrees the nucleus lies off the axial axis of symmetry. However, these relations are only approximate because the complex is not truly axially symmetric and a second (contact) shift mechanism may also be involved. The physical properties of the donor–La(dkt)$_3$ complexes (dkt = diketone) have been thoroughly studied. Rates of complex formation and dissociation are usually high on the NMR time scale, although a few complexes have been observed in the slow-exchange limit at low temperature (Evans and Wyatt, 1972, 1973; Grotens et al., 1973). The stoichiometry of the complexes can be quite variable, 1 : 1, 2 : 1, or 3 : 2 (Horrocks and Sipe, 1971; Evans and Wyatt, 1972, 1973, 1974; Shapiro and Johnston, 1972; Cunningham and Sievers, 1975), and more than one type may be formed between a given donor and the shift reagent. In addition, the conformation of the organic donor may be altered in each complex so that application of the McConnell equation is fraught with complexities. This has not deterred the use of shift reagents to determine precise conformational properties of the donors, and a very thorough and critical assessment of this method as a conformational probe has appeared (Inagaki and Miyazawa, 1980). Fortunately, these physical properties of the shift reagent–donor complexes are of peripheral interest to the topic at hand. The major application of CSRs depends only on their capacity to effect different induced shifts[1] for protons in different stereochemical environments so that integration of the resolved signals will provide a direct measure of enantiomer ratios.

---

[1]The downfield shifts caused by Eu(dpm)$_3$ have also been referred to as pseudocontact shifts and bound shifts. Since both the cause of the shift and the stoichiometry vary, it is more correct to use the term *induced shift* to represent $\Delta\delta$.

This elegant achievement was first described by Whitesides and Lewis (1970), who synthesized the chiral ligand 3-pivaloyl-*d*-camphor (pvc) and its europium complex, $Eu(pvc)_3$. This shift reagent exhibited the capacity to induce separation of the proton resonances for the enantiomers of many common organic donors such as α-phenylethylamine and α-phenylethanol. Figure 2 shows the well-resolved signals for the methyl, methine, and ortho aromatic portions of α-phenylethylamine in the presence of 0.5 equivalent of $Eu(pvc)_3$. The separate

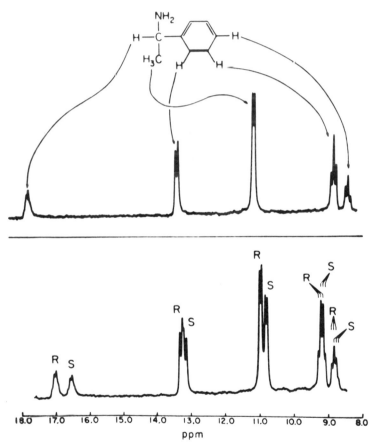

Fig. 2. 100-MHz ^{1}H-NMR spectra of a solution prepared from (upper) (*S*)-α-phenethylamine (10 μl) and (lower) a mixture of (*R*)- and (*S*)-α-phenethylamine (7 and 5μl, respectively) in 0.3 ml CCl_4 containing $Eu(pvc)_3$ (0.15 *M*). The δ scale applies to the mixture only. The upper trace is slightly displaced to lower field due to the use of a slightly different concentration. From Whitesides and Lewis (1970). Copyright 1970 by the American Chemical Society.

signals for the *R*- and *S*-enantiomers result from induced shifts of different magnitudes and are reported as $\Delta\delta$ in units of parts per million. Before the practical accomplishments of this and other CSRs are described, it is of value to examine the stereochemical properties of the CSRs themselves.

III. Structural Isomerism in Chiral Shift Reagents

Every hexacoordinate lanthanide complex of a symmetric diketone is chiral and can exist in the two enantiomeric forms shown in Fig. 3. All accumulated evidence indicates that the interconversion of these two forms, the Λ and Δ isomers, are rapid on the chemical time scale (Sievers, 1973; McCreary *et al.*, 1974) for all lanthanide trisdiketonates. If the diketone is not symmetrically substituted (R $\neq$ R') there will also be a pair of cis- and trans-isomers in both series. Figure 4 depicts the stereochemical arrangements in the cis- and trans-Δ-isomers. These, too, would be expected to interconvert readily. Indeed such an intramolecular rearrangement of either Eu(fod)$_3$ or Pr(fod)$_3$ (fod = heptafluoro-propanoylpivaloylmethanato ligand) was too rapid for observation by ^{1}H-NMR spectroscopy at $-100°C$ (Evans and deVillardi 1977, 1978). Only when the metal ion in the octahedral complex is a transition element is the interconversion slow (*a*) on the NMR time scale for V^{3+} and (*b*) on the chemical time scale for Co^{3+} (Everett and Chen, 1970).

When the diketone is itself chiral as in pivaloyl-*d*-camphor, the CSR derived therefrom consists of four rapidly interconverting chiral diastereomers (cis-Λ, cis-Δ, trans-Λ, and trans-Δ.) For each cis-isomer there will be four different donor sites resulting in four different CSR–donor complexes. These are shown as S^1, S^2, S^3, and S^4 in Fig. 5. For each trans-isomer two isomeric complexes can be formed. Still more complications arise from additional isomers for the

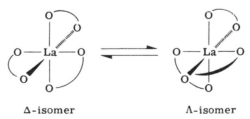

Δ-isomer Λ-isomer

Fig. 3. Three-dimensional representation of the enantiomeric forms of a lanthanum complex of a symmetric diketone.

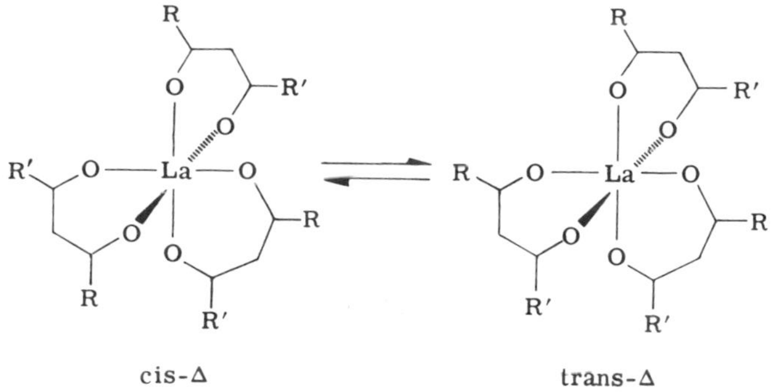

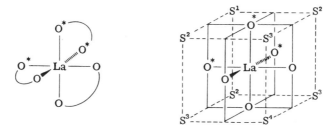

Fig. 4. Three-dimensional representation of two of the four diastereomeric complexes formed from combination of a La^{3+} ion with 3 equivalents of an unsymmetric diketone.

$CSR-(donor)_2$ species as has been described (Horrocks, 1973). Furthermore, shift reagents associate as dimers in carbon tetrachloride (Brittain and Richardson, 1976) and these dimers may also interact with the donors. We have noticed on several occasions that the solution of an organic donor and CSR exhibits changes in its NMR spectrum for the first 10 min after preparation, possibly due to this slow attainment of all equilibrium processes including equilibration of all possible donor–$La(dkt)_3$ complexes and of monomeric and dimeric shift reagent species (Evans and deVillardi, 1978). There is also evidence that the structure of the shift reagent itself changes upon complexation, as is indicated by changes in the optical circularly polarized emission (CPE) spectra when the donor is added (Brittain and Richardson, 1976; Brittain, 1980).

We are fortunate that a detailed knowledge of all isomeric CSR–donor structures is not necessary for their proper use, but their complexity should be kept in mind in the interpretation of experimental results.

Fig. 5. Three-dimensional representation of the four sites (S^1–S^4) presented to a donor by the cis-Δ-isomer of a shift reagent.

IV. Variations in Structure

A. Diketone

The most effective ligands for use in CSRs are the acyl camphors (**1–3**), dicampholylmethane (**4**), and trifluoroacetylnopinane (**5**). Compounds **1**, **4**, and **5** were synthesized by McCreary *et al.* (1974) along with a large number of other ligands that proved to be less effective. Compounds **2** and **3** were synthesized by the groups of Goering (Goering *et al.,* 1971) and Fraser, respectively (Fraser *et al.,* 1971). The europium, praseodymium, and ytterbium diketonates of **2** and **3** are all commercially available, as is the europium chelate of **4**. Table I lists the commercial sources of the CSRs derived from **2**, **3**, and **4** and the common abbreviations by which they are most frequently described.

TABLE I
Commercially Available Chiral Shift Reagents

Structure	Metal	Abbreviation[a]	Suppliers[b]
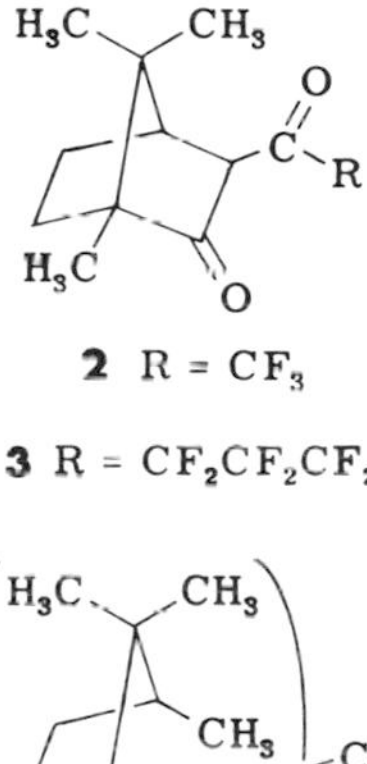 **2** R = CF$_3$	Eu	Eu(tfc)$_3$, Eu-2	A, B, D–G
	Pr	Pr(tfc)$_3$, Pr-2	A, B, D–G
	Yb	Yb(tfc)$_3$, Yb-2	B, G
3 R = CF$_2$CF$_2$CF$_2$	Eu	Eu(hfbc)$_3$, Eu-3	A, C, D, G
	Pr	Pr(hfbc)$_3$, Pr-3	A, D, G
	Yb	Yb(hfbc)$_3$, Yb-3	G
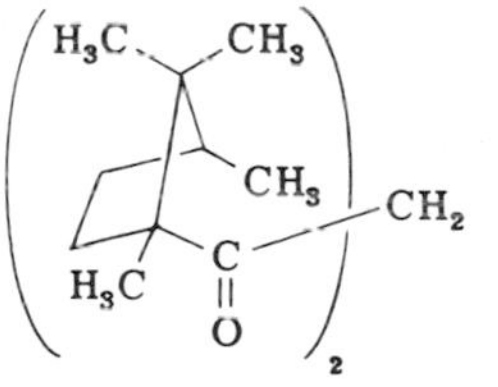 **4**	Eu	Eu(dcm)$_3$, Eu-4	B

[a] tfc, Trifluorohydroxymethylene-*d*-camphorato (also called facam); hfbc, hepta-fluoropropylhydroxymethylene-*d*-camphorato; dcm, dicampholyl-*d*-methanato.

[b] A, Aldrich Chemical Co., Milwaukee, Wisconsin; B, Alfa Division, Ventron Corp., Danvers, Massachusetts; C, Chemical Dynamics Corp., South Plainsfield, New Jersey; D, Norell Inc., Landsville, New Jersey; E, Regis Chemical, Morton Grove, Illinois; F, Sigma Chemical Co., St. Louis, Missouri; G, Stohler Isotopic Chemicals, Waltham, Massachusetts.

$1\ \text{R} = \text{C(CH}_3)_3$
$2\ \text{R} = \text{CF}_3$
$3\ \text{R} = \text{CF}_2\text{CF}_2\text{CF}_3$

4

5

Comparisons to determine the relative resolving power of the various europium and praseodymium diketonates of 1–4 have been relatively limited and have often been made using different shift reagent–donor ratios and at different concentrations. This makes comparisons between the different sets of data difficult because the effect of the concentration of shift reagent on $\Delta\delta\delta$ is not usually linear.

Table II describes the effects on the proton shifts of 1-phenylethanol, 1-phenylethylamine, and benzylmethyl sulfoxide as a result of adding Eu-2, Eu-3, Eu-4, and Eu-5. Also given for comparison are the effects of Eu-3 and Pr-3 on the ^{13}C signals of the same alcohol and amine. Since the effects of **1** are consistently weaker, they are not included. It is clear that the best resolving power among the europium chelates is shown by Eu-4, the other three being comparable on the average. It can also be seen that the effects of Eu-3 are much larger in the ^{13}C than in the ^{1}H spectra, particularly if it is noted that only one-fifth as much reagent was used in the ^{13}C study.

B. Lanthanide Atom

Table II also shows that Pr-3 is significantly better than Eu-3, indeed approaching Eu-4 in its resolving power. That Pr-3 has better resolving power than Eu-3 is most strikingly demonstrated by its capacity to induce nonequivalence in the benzylic protons of benzyl alcohol and the methyl groups of dimethyl sulfoxide (Fraser et al., 1972) at one-fifth the concentration of Eu-3 necessary for the same effect (Kainosho et al., 1972). For each racemate the order of resolving power was Pr-3 > Eu-3 > Eu-2, with Pr-3 being 5–10 times stronger than Eu-2. There are many indications from studies on achiral shift reagents that a third lanthanide element, ytterbium, can give reliably large induced shifts with little line broadening. Thus, the commercially available Yb-2 and Yb-3 may prove to be among the best in resolving power. Only one article has made a comparison of Yb-3 with Pr-3. The authors (Tangerman and Zwanenburg, 1977) found Yb-3 to be superior to Pr-3 in studies on a variety of sulfoxide derivatives. The conclu-

TABLE II
Enantiomeric Shift Differences for Chiral Shift Reagents[a]

		Protons enantiotopic by external comparison			
Compound	Signal	Eu-2	Eu-3	Eu-4	Eu-5
$C_6H_5CHNH_2$	CH_3	0.5 (0.18)	(0.17) [0.08]	0.7	0.21
|	CH	(0.92)	(0.22) [0.00]	4.4	0.10
CH_3	H_0	0.05	[0.06]	0.15	0.12
C_6H_5CHOH	CH_3	0.0 (0.0)	(0.22) [0.05]	0.61	0.55
|	CH	0.4 (0.37)	(0.40) [0.07]	0.7	0.45
CH_3	H_0	0.0	[0.02]	0.06	0.05
O ↑ $C_6H_5CH_2SCH_3$	CH_3	0.0	[0.06]	1.21	0.05

		Protons enantiotopic by internal comparison			
Compound	Signal	Yb-2	Eu-2	Eu-3	Pr-3
O ↑ CH_3SCH_3	CH_3	|0.16|	(0.0) |0.06|	(0.1)	[0.17]
$C_6H_5CH_2OH$	CH_2	—	—	(0.28)	[0.26]

	Carbons enantiotopic by external comprison[b]		
Compound	Signal	Eu-3	Pr-3
C_6H_5CHOH	CH_3	0.0	0.0
|	CH	0.11	0.28
CH_3	C_1	0.0	0.17
$C_6H_5CHNH_2$	CH_3	0.18	0.0
| CH_3	CH	0.34	0.2
$C_6H_5CH_2SOCH_3$	CH_3	0.0	0.15
	CH_2	0.0	0.0

[a]The data that are not enclosed in parentheses or brackets were obtained using 0.3 M substrate and 0.3 M CSR (McCreary et al., 1974). The data in parentheses were also obtained at 0.3 M concentration of both (Goering et al., 1974). The data in brackets were obtained with 0.3 M substrate and only 0.12 M shift reagent (Fraser et al., 1971). The figures between vertical bars were obtained using 0.1 M substrate and 0.05 M shift reagent (Kainosho et al., 1972).
[b]The ^{13}C shift differences were determined using 0.6 M substrate and 0.12 M shift reagent (Fraser et al., 1973).

sions arrived at in separate studies (McCreary *et al.*, 1974; Goering *et al.*, 1974; Fraser *et al.*, 1972) are all in agreement in the sense that no one CSR proved to be superior with all donors. All recommended that several reagents be tried in any situation. On the basis of all the evidence, the orders of preference would be Yb ≈ Pr > Eu and **4** > **3** > **2**. It is unfortunate that Pr-**4** and Yb-**4** are not yet commercially available. One additional factor should be kept in mind when one is choosing the appropriate CSR. In contrast to the downfield shifts induced by europium and ytterbium complexes, those of praseodymium are to high field. In certain cases this may be an adverse property in that many peaks will interfere with one another in the region of δ 0–3. At other times upfield shifts may prove to be advantageous as, for example, when one is observing methyl peaks.

V. Mechanism of Action

To aid in the selection of a CSR appropriate to a given task and in the interpretation of its effects under optimal experimental conditions, it is important to be aware of the various mechanisms by which a CSR achieves its resolution. Whitesides has proposed the recognition of two mechanisms by which the resolution of signals for two enantiomers can be accomplished (McCreary *et al.*, 1974). (*a*) A CSR and a racemix mixture may form two "diastereomeric complexes," which will have different stabilities reflected in the values for their dissociation constants; as a result, all the signals for one diastereomer will be more greatly affected than are those for the other more weakly bound isomer. (*b*) Each diastereomer will of necessity form complexes having different geometries; this will, through the angular and distance factors in the McConnell equation, produce different induced shifts. A variety of evidence (McCreary *et al.*, 1974) indicates that both mechanisms are to be operative, but the extent of the individual contributions cannot be evaluated. The fact that each 1 : 1 complex may be a mixture of at least 12 isomers and that an equal number of 2 : 1 isomers may be formed as well makes this evaluation impossible.[2]

That the second mechanism is of considerable importance is indicated by the successful resolution of signals for nuclei that are enantiotopic by internal comparison.[3] As mentioned in Section IV,B, the methylene protons of benzyl alco-

[2]The viability of (*a*),preferential complexation, is clearly established by the report of the resolution of propylene oxide by passing through a column of Eu-**2** or Pr-**3** on Chromosorb (Golding *et al.*, 1977).

[3]Mislow has defined two classes of enantiotopic nuclei: those enantiotopic by external comparison, as exemplified by the two methine protons in the two enantiomers of RCHOHR′, and those enantiotopic by internal comparison such as the prochiral methylene protons of RCH₂R′ (Raban and Mislow, 1967).

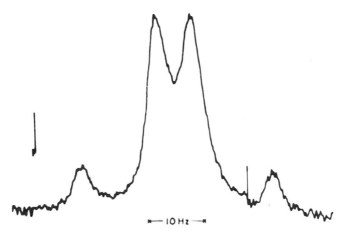

Fig. 6. The 100-MHz ^{1}H-NMR spectrum of the methylene proton absorption of a 0.2 M solution of benzyl alcohol in CCl_4 containing 0.15 equivalent of Pr-3. From Fraser *et al.* (1972). Copyright 1972 by the American Chemical Society.

hol become diastereotopic in the presence of a CSR. Thus, the nonequivalent CH_2 absorption for the CH_2 group of the CSR–ϕCH_2OH complex appears as the well-separated AB quartet shown in Fig. 6. This nonequivalence of 0.26 ppm was achieved using 0.4 equivalent of Pr-3. Since 2 M Eu-3 was required for the same effect on benzyl alcohol but at the 0.4 M level for nuclei that are enantiotopic by external comparison, it seems that factor (a) is usually greater. The single conclusion that has been drawn from all the data on CSRs is that the magnitude of $\Delta\delta\delta$ does not exhibit consistent behavior with respect either to concentration or to temperature and thus most likely reflects a complex composite of mechanisms.

VI. Uses of Chiral Shift Reagents

A. *Determination of Enantiomer Ratios*

1. Strong Organic Bases

As the spectra in Fig. 2 show, many of the ^{1}H signals for a pair of enantiomers are sufficiently separated by the presence of a CSR to permit one to make an accurate determination of their ratio by integration. By far the most common nucleus studied has been ^{1}H, yet other readily measurable nuclei such as ^{13}C, ^{19}F, and ^{31}P have proved useful, and, of course, any magnetically active nucleus

TABLE III

Representative Examples of Enantiomer Ratio Measurements

Class	Structure	Shift reagent	Reference
Alcohols			
Primary	ϕCHDCD$_2$OH	Eu-**4**	Lau et al. (1976)
Secondary	ϕCH(OH)C≡CR	Eu-**4**	Midland et al. (1980)
	CH$_3$CH(OH)CH$_2$COOC$_2$H$_5$	Eu-**3**	Meyers and Amos (1980); Meyers and Yamamoto (1981)
Tertiary	CH$_3$C(OH)(CH$_2$)$_3$CH$_3$ | COOCH$_3$	Eu-**2**	Frater et al. (1981); cf. Eliel and Frazee (1979)
Diols	RCH(OH)CH$_2$CHR′CH$_2$OH	Eu-**3**	Jakovac and Jones (1979)
Ethers			
Biaryl	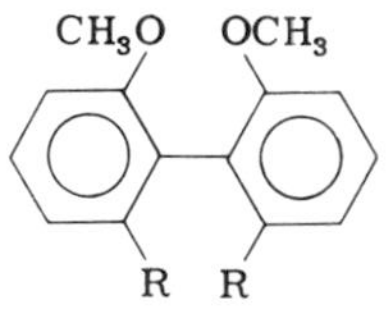	Eu-**3**	Diaz et al. (1981); Behnam et al. (1979)
Cyclic	ϕCH—CH$_2$ \\ / O	Eu-**3**, Pr-**3**	Fraser et al. (1973)
Methoxy ester	RCH=CH—CH=CH | CH$_3$CH(OCH$_3$)COO	Eu-**3**	Trost et al. (1980)
Aldehydes			
Aliphatic	ϕCH(CH$_3$)CHO	Eu-**3**	Fraser et al. (1971)

184

Ketones			
Acyclic	$\phi CH_2COCH(CH_3)\phi$	Eu-3	Enders and Eichenauer (1979)
Cyclic	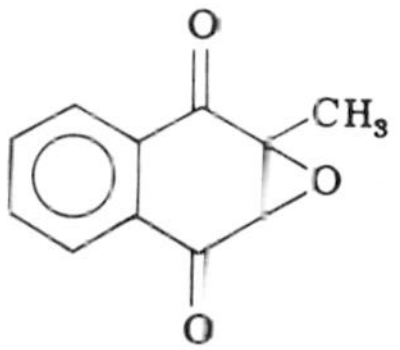	Eu-4	Pluim and Wynberg (1980)
	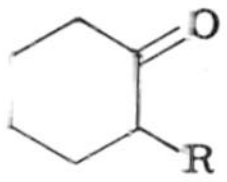	Eu-3	Meyers et al. (1981a)
	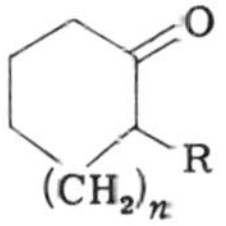 $n = 5, 7, 10$	Eu-3	Meyers et al. (1981b)
	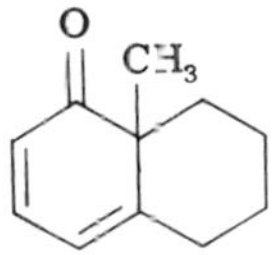	Eu-2	Dugan and Murphy (1976)
	Bianthrones	Eu-3	Agranat et al. (1979)
	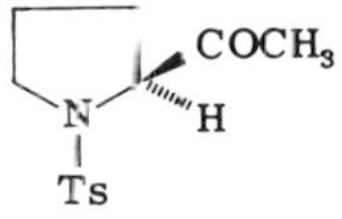	Eu-3	Evans et al. (1981)

(continued)

TABLE III (*Continued*)

Class	Structure	Shift reagent	Reference
Esters			
Aliphatic	$(CH_3)_2C\!=\!CH(CH_2)_2CH(CF_3)CH_2COOCH_3$	Eu-4	Valentine *et al.* (1980)
	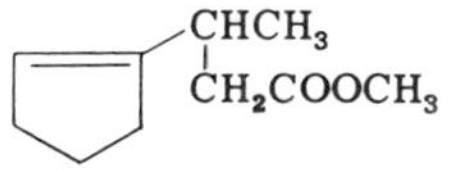 (structure) CH_2COOCH_3, $CHCH_3$	Eu-4	Trost and Strege (1977)
	$CH_3CH\!=\!C\!=\!C(CH_3)COOCH_3$	Eu-4	Lang and Hansen (1979a)
Difunctional	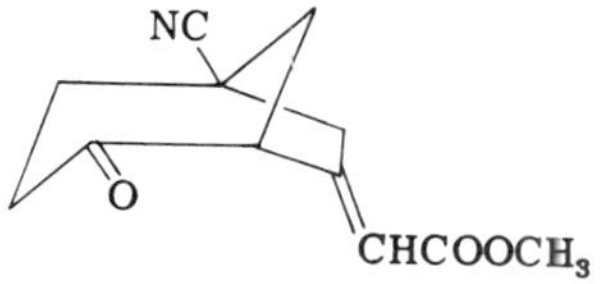 (bicyclic structure) NC, CHCOOCH$_3$, O	Eu-4	Trost *et al.* (1979)
Amines			
Primary	$\phi CH(CH_3)NH_2$	Eu-1	Whitesides and Lewis (1970)
	$(CH_3)_3CCH(NH_2)COOC(CH_3)_3$	Eu-2	Hashimoto *et al.* (1976)
Secondary	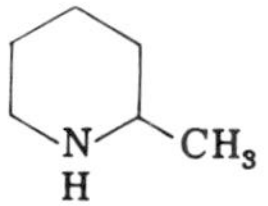 (piperidine) N–H, CH$_3$	Eu-3, Pr-3	Fraser *et al.* (1973)
	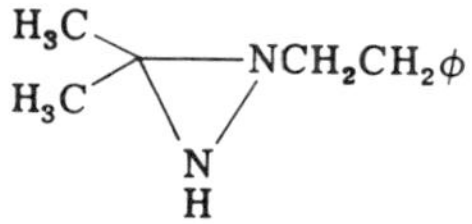 (aziridine) H$_3$C, H$_3$C, NCH$_2$CH$_2\phi$, N–H	Eu-3	Mintas *et al.* (1981)

186

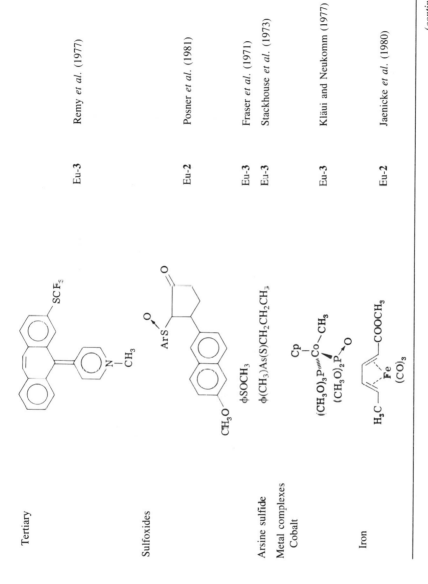

Tertiary		Eu-3	Remy et al. (1977)
Sulfoxides		Eu-2	Posner et al. (1981)
Arsine sulfide	φSOCH₃	Eu-3	Fraser et al. (1971)
	φ(CH₃)As(S)CH₂CH₂CH₃	Eu-3	Stackhouse et al. (1973)
Metal complexes Cobalt		Eu-3	Kläui and Neukomm (1977)
Iron		Eu-2	Jaenicke et al. (1980)

$\phi SOCH_3$

$\phi(CH_3)As(S)CH_2CH_2CH_3$

(continued)

187

TABLE III (*Continued*)

Class	Structure	Shift reagent	Reference
	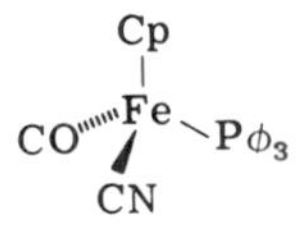 Cp CO⎞⎞⎞Fe—Pφ₃ CN	Yb-**2**	Reger (1975)
Platinum			
	Cp H₃C⎞⎞⎞Pt—COCH₃ H₅C₂	Eu-**2**	Hamer and Shaver (1980)
Natural products and drugs			
Amino acids		Eu-**3**	Schollköpf *et al.* (1981)
Mevalolactone		Eu-**3**	Eliel and Soai (1981)
Multistriatin		Eu-**2**	Mori (1975)
Frontalin		Eu-**2**	Mori (1976)
Penicillamine (acetonide)		Eu-**3**	Cockerill *et al.* (1974)
Pavines and other *N*-alkaloids		Eu-**2**	Shaath and Soine (1975)
Prostaglandins		Eu-**3**	Morton and Morge (1978)
		Eu-**3**	Bus and Korver (1975)

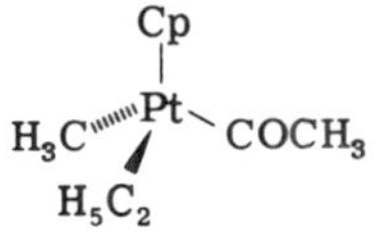

could serve as a potential probe. Most of the chiral donors whose enantiomer signals have been readily resolved belong to the category of "hard bases" because the lanthanide shift reagents are "hard Lewis acids." Table III presents a list of representative organic bases to which application of the CSR method of measuring enantiomer ratios has been successful. From these data it should be strikingly clear that potentially the best CSRs, La-**4**, Pr-**3**, and Yb-**3**, have only rarely been used by organic chemists.

The obvious advantage of a successful CSR measurement is that the result is unambiguous and the accuracy of the enantiomeric excess thereby obtained is high (proton integrals can normally be accurate to $\pm 2\%$). The only disadvantage is that its success is not universal.

The effect of temperature on the resolving power of a CSR, although neither widely studied nor widely recognized, can be very large and beneficial. For example, the enhancement in $\Delta\delta$ for the proton signals of styrene oxide in the presence of Eu-**3** as the temperature is lowered is described by the data in Table IV. It is noteworthy that a 50° drop in the temperature (T) caused a two- to fivefold increase in resolution, depending on the signal studied. It is also significant that one of the proton signals (H_0) actually lost resolution as T decreased. Because in theory four effects contribute to the increase in the induced shifts of donor signals and these are operative on an indeterminate mixture of diastereomeric complexes, it should not be surprising to find such variations in $\Delta\delta$ with a change in T. What is important is the recognition of the potential benefit of this simple experimental parameter by users.

Thus, if one wishes to determine the enantiomeric excess of an unknown sample, the following sequence is recommended. At least four CSRs should be tried, the approximate order of potential capacity being Eu **4** > Pr-**3** ≈ Yb-**3** > Eu-**3**. This seems advisable because there is ample evidence from many sources (McCreary *et al.*, 1974; Goering *et al.*, 1974; Hlubecek and Shapiro, 1972;

TABLE IV

Effect of Temperature on Downfield Shift ($\Delta\delta$)
and Enantiomeric Shift Differences ($\Delta\Delta\delta$)[a]
Using 0.09 M Eu-3 and 0.2 M Styrene Oxide

$$\phi \overset{}{\underset{HC}{\diagdown}} C - C \overset{H_A}{\underset{H_B}{\diagup}}$$
$$O$$

Temp (°C)	H_A	H_B	H_C	H_0
41	4.22 (0.17)	4.33 (0.64)	4.75 (0.09)	2.30 (0.02)
27	4.79 (0.22)	4.90 (0.88)	5.36 (0.22)	2.67 (0.00)
−6	6.21 (0.39)	6.29 (1.65)	6.79 (0.50)	3.59 (0.00)

[a] $\Delta\Delta\delta$ volumes are in parentheses. All values are expressed as parts per million.

Welti *et al.*, 1978) that the geometry of any shift reagent varies with the lanthanide metal, as does its resolving power, in an unpredictable way. If success is
still lacking, then one should study the solutions at higher and lower temperatures. As a last resort one may be able to convert the unknown to a derivative that
will be more basic (e.g., phosphine to a phosphine oxide) and study it in the
presence of CSRs.

2. WEAK ORGANIC BASES

Very small and often negligible are the shifts induced by the addition of a shift
reagent to olefinic, aromatic, and halogen compounds. Nevertheless, induced
shifts have been created or significantly increased by the addition of silver
perfluorobutyrate (Evans *et al.*, 1975). Further improvement was attained by
Sievers and co-workers (Wenzel *et al.*, 1980, Wenzel and Sievers, 1981), who
were able to induce shifts of several parts per million in the above classes of
compounds by using Pr(fod)$_3$ or Yb(fod)$_3$ along with Ag(fod) as an additive. It
was shown (Offermann and Mannschreck, 1981) that by the use of the CSRs
Eu-**3**, Pr-**3**, and Yb-**3** in conjunction with Ag(fod) excellent resolution of both
the ^{1}H and ^{13}C resonances of the four chiral olefines **6–9** could be achieved. The
structures of the compounds studied (**6–9**) are shown with the $\Delta\delta$ values listed
beside the nucleus studied. The underlined values for **6** and **7** represent ^{13}C
signals; others represent protons.

These observations of Offermann and Mannschreck have been corroborated
(Wenzel and Sievers, 1982) in an article on the effects of "lanthanide chelate–silver(I) binuclear" shift reagents.

3. IN AQUEOUS MEDIUM

Although Whitesides and Lewis had established the capacity of a chiral donor to induce chirality in an otherwise achiral shift reagent,[4] the utilization of this principle in aqueous solutions of lanthanide ions was only fairly recently established (Reuben, 1980). Reuben showed that racemic mixtures of several α-hydroxycarboxylates give rise to resolvable methyl group signals (*a*) in the presence of $EuCl_3$ or $PrCl_3$ and (*b*) in solutions containing both lanthanide halide and one enantiomer of malate or citramalate (3 eqivalents). The latter conditions should prove to be of practical use for determining the enantiomeric excess of α-hydroxy acids in aqueous solution. Reuben also succeeded in resolving signals that were enantiotopic by internal comparison, specifically the (*pro-R*)- and (*pro-S*)-hydrogens of glycolate, $HOCH_2COO^-$, in the presence of 1.2 equivalents of $PrCl_3$ and 5 equivalents of *l*-lactate (Reuben, 1980).

B. Assignment of Configuration to Enantiomers

There have been several investigations of the possibility of arriving at absolute configurational assignments on the basis of the relative magnitudes of $\Delta\delta\delta$ (Dongala *et al.*, 1972; Reich *et al.*, 1973; Sullivan *et al.*, 1974). In view of a failure in one closely related series (Sullivan *et al*, 1974) and the many factors potentially responsible for enantiomeric shift differences, it seems generally accepted that any assignments of absolute configuration from CSR data would be ambiguous.

C. Assignment of Configuration to Meso- versus dl-Isomers

Since meso- and *dl*-pairs, as diastereoisomers, always interact in different ways with a CSR, the manifestation of their differences frequently allows a structural assignment to be made, although the basis for distinction will vary with the type of prochirality present in the meso-isomer.

For example, the methyl and the methine protons of *meso*-2,3-butene oxide are enantiotopic by internal comparison. As a result, in the presence of a CSR both sets of protons become diastereotopic and will exhibit different chemical shifts, producing an A_3BCD_3 spin system. The distinguishing feature will be the appearance of the coupling between the methine protons in the shifted spectrum. In contrast, the methine protons of the racemate, being enantiotopic by external comparison, will appear as two racemate patterns similar to that of the original

---

[4]In a related experiment (Pirkle and Sikkenga, 1975) it was found that $Eu(fod)_3$ in a chiral solvent accentuated the resolving power of that solvent.

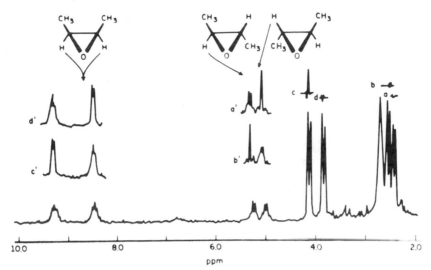

Fig. 7. 100-MHz ¹H-NMR spectra of a mixture of *meso-* and (*dl*)-2,3-butylene oxide in the presence of Eu-2. The inset double-resonance spectra were obtained by irradiating at a and observing decoupling at a', etc. From Kainosho *et al.* (1972). Copyright 1972 by the American Chemical Society.

racemate, that is, $A_3BB'A_3'$ systems. This method, successfully applied to the butene oxides (Kainosho *et al.*, 1972), produced the spectra shown in Fig. 7. In a similar way the configuration of the isomeric 2,3-difluorosuccinates were readily assigned on the basis of the absorption patterns observed in the presence of a CSR for the meso-(ABXY) and the *dl*-isomers (AA'BB') (Bell and Hudlicky, 1980). A different type of behavior provided a distinction between the two isomeric diols **10** and **11** (Johnson *et al.*, 1975). The racemate **10** was identified as the isomer that gave rise to two *tert*-butyl singlets in the presence of a CSR. The method has also been successfully applied to the assignment of structure to a pair of iron complexes of dimethyl fumarate and succinate (Schurig, 1976) and to nickel complexes of *meso-* and *dl*-2,3-diaminobutane derivatives (Lindoy and Moody, 1977).

D. Measurement of Rotational Barriers

The variable-temperature NMR method (DNMR) has been widely used to obtain activation parameters for barriers to rotation or inversion in a wide variety of organic compounds (Jackman and Cotton, 1975). The success of the method depends on the capacity of NMR to resolve the signals for nuclei in two different environments. For example, the protons of the two methyl groups in the chiral nitrosamine **12** are diastereotopic, but in the nitrosamine **13** they are enantiotopic.

12   **13**

Mannschreck and co-workers (1973) were the first to point out that in achiral molecules such as **13** the barrier to rotation about the Ar—N bond interconverts the enantiotopic methyls. It therefore can be measured by studying the DNMR behavior of **13** in the presence of a CSR. They later showed that the shift reagent does not itself perturb the magnitude of the barrier at the requisite concentration (Lefèvre *et al.*, 1977). A similar conclusion was arrived at independently (Tangerman and Zwanenburg, 1977) from studies on a series of thiocarbonyl compounds using prochiral methyl groups as probes. The method has also been applied to chiral butadienes (Mannschreck *et al.*, 1974; Becher *et al.*, 1978; Holik and Mannschreck, 1979a) and pivalophenones (Holik and Mannschreck, 1979b).

This use of a CSR does not seem to have any major advantage other than to permit the study of structurally simpler systems.

E. Measurement of Coupling Constants

As discussed earlier using the example shown in Fig. 6, nuclei that are enantiotopic by internal comparison can be induced to exhibit resolvable shift differences in the presence of a CSR. As a result the geminal coupling constant between these protons can be measured. By this method Lang and Hansen (1979b) were able to observe the geminal coupling constants between the allenic methylene protons of **14** in the presence of Pr-3.

14

A second application of the influence of Pr-**3** on the protons of benzyl alcohol was to check the enantiomeric purity of (S)-C_6H_5CHDOH, which had been obtained by the microbiological reduction of [^{2}H]benzaldehyde (Nagai and Kobayashi, 1976).[5]

VII. Experimental Procedures

A number of detailed procedures for the preparation of lanthanide trisdiketonates have appeared. The reaction can be carried out in ethanol–water (Eisentraut and Sievers, 1965) or methanol (McCreary et al., 1974). The barium salt of the diketone can also be used (Schurig, 1972). Two aspects of the general method of preparation deserve special comment. First, a serious difficulty will be encountered unless the sodium hydroxide or alkoxide solutions have been freshly prepared. For complete protection against carbon dioxide, all preparations should be carried out under an inert atmosphere. In our experience it is also advantageous to sublime the sample, to ensure the removal of water, although others (Goering et al., 1974; McCreary et al., 1974) use the product after drying in vacuo. Sublimation temperatures are generally about 200°C if a vacuum of 0.05 mm or better is employed.

References

Agranat, I., Tapuhi, Y., and Lallemand, J. Y. (1979). Nouv. J. Chim. **3**, 59.

Becher, G., Burgemeister, T., Henschel, H. H., and Mannschreck, A. (1978). Org. Magn. Reson. **11**, 481.

Behnam, B. A., Hall, D. M., and Modara, B. (1979). Tetrahedron Lett., p. 2619.

Bell, H. M., and Hudlicky, M. (1980). J. Fluorine Chem. **15**, 191.

Brittain, H. G. (1980). Inorg. Chem. **19**, 2233.

Brittain, H. G., and Richardson, F. S. (1976). J. Am. Chem. Soc. **98**, 5858.

Bus, J., and Korver, O. (1975). Recl. Trav. Chim. Pays-Bas **94**, 254.

Cockerill, A. F., Davies, G. L. O., Harrison, R. G., and Rackham, D. M. (1974). Org. Magn. Reson. **6**, 669.

Cunningham, J. A., and Sievers, R. E. (1975). J. Am. Chem. Soc. **97**, 1586.

Diaz, E., Rojas-Dávilla, E., Guzmán, A., and Joseph-Nathan, P. (1980). Org. Magn. Reson. **14**, 439.

Dongala, E. B., Solladié-Cavallo, A., and Solladié, G. (1972). Tetrahedron Lett., p. 4233.

[5]Added in proof: In addition to the analytical applications for shift reagents, a synthetic use has been reported by M. Bednarski and S. Danishefsky [(1983) J. Am. Chem. Soc. **105**, 3716]. Catalysis of a Diels–Alder reaction by Eu-**3** gave an adduct with an enantiomeric excess of 50%.

Duggan, P. G., and Murphy, W. S. (1976). *J. Chem. Soc. Perkin Trans.* **1**, p. 634.

Eisentraut, K. J., and Sievers, R. E. (1965). *J. Am. Chem. Soc.* **87**, 5254.

Eliel, E. L., and Frazee, W. J. (1979). *J. Org. Chem.* **44**, 3598.

Eliel, E. L., and Soai, K. (1981). *Tetrahedron Lett.*, p. 2589.

Enders, D., and Eichenauer, H. (1979). *Angew. Chem. Int. Ed. Engl.* **18**, 397.

Evans, D. A., Nelson, J. V., Vogel, E., and Taber, T. R. (1981). *J. Am. Chem. Soc.* **103**, 3099.

Evans, D. F., and deVillardi, G. C. (1977). *J. Chem. Soc. Dalton Trans.*, p. 2256.

Evans, D. F., and deVillardi, G. C. (1978). *J. Chem. Soc. Dalton Trans.*, p. 315.

Evans, D. F., and Wyatt, M. (1972). *J. Chem. Soc. Chem. Commun.* p. 312.

Evans, D. F., and Wyatt, M. (1973). *J. Chem. Soc. Chem. Commun.*, p. 339.

Evans, D. F., and Wyatt, M. (1974). *J. Chem. Soc. Dalton Trans.*, p. 765.

Evans, D. F., Tucker, J. N., and de Villardi, G. C. (1975). *J. Chem. Soc. Chem. Commun.*, p. 205.

Everett, G. W., Jr., and Chen, Y. T. (1970). *J. Am. Chem. Soc.* **92**, 508.

Fraser, R. R., Petit, M. A., and Saunders, J. K. (1971). *J. Chem. Soc. Chem. Commun.*, p. 1450.

Fraser, R. R., Petit, M. A., and Miskow, M. (1972). *J. Am. Chem. Soc.* **94**, 3253.

Fraser, R. R., Stothers, J. B., and Tann, C. T. (1973). *J. Magn. Reson.* **10**, 95.

Frater, G., Müller, U., and Günther, W. (1981). *Tetrahedron Lett.*, p. 4221.

Goering, H. L., Eikenberry, J. N., and Koermer, G. S. (1971). *J. Am. Chem. Soc.* **93**, 5913.

Goering, H. L., Eikenberry, J. N., Koermer, G. S., and Lattimer, C. J. (1974). *J. Am. Chem. Soc.* **96**, 1493.

Golding, B. T., Sellars, P. J., and Wong, A. K. (1977). *J. Chem. Soc. Chem. Commun.*, p. 571.

Grotens, A. M., Backus, J. J. M., Pijpers, F. W., and de Boer, E. (1973). *Tetrahedron Lett.*, p. 1467.

Hamer, G., and Shaver, A. (1980). *Can. J. Chem.* **58**, 2011.

Hashimoto, S., Yamada, S., and Koga, K. (1976). *J. Am. Chem. Soc.* **98**, 7450.

Hinckley, C. C. (1969). *J. Am. Chem. Soc.* **91**, 5160.

Hlubucek, J. R., and Shapiro, B. L. (1972). *Org. Magn. Reson.* **4**, 825.

Hofer, O. (1976). *Top. Stereochem.* **9**, 111.

Holik, M., and Mannschreck, A. (1979a). *Org. Magn. Reson.* **12**, 223.

Holik, M., and Mannschreck, A. (1979b). *Org. Magn. Reson.* **12**, 28.

Horrocks, W. D., Jr. (1973). *In* "NMR of Paramagnetic Molecules" (G. N. LaMar, W. D. Horrocks, and R. H. Holm, eds.), p. 479. Academic Press, New York.

Horrocks, W. D., Jr., and Sipe, J. P., III (1971). *J. Am. Chem. Soc.* **93**, 6800.

Inagaki, F., and Miyazawa, T. (1980). *Prog. NMR Spectrosc.* **14**, 67.

Jackman, L. M., and Cotton, F. A. (1975). "Dynamic Nuclear Magnetic Resonance Spectroscopy." Academic Press, New York.

Jaenicke, O., Kerber, R. C., Kirsch, P., Koerner von Gustorf, E. A., and Rumin, R. (1980). *J. Organomet. Chem.* **187**, 361.

Jakovac, I. J., and Jones, J. B. (1979). *J. Org. Chem.* **44**, 2165.

Johnson, P. Y., Jacobs, I., and Kerkman, D. J. (1975). *J. Org. Chem.* **40**, 2710.

Kainosho, M., Ajisaka, K., Pirkle, W. H., and Beare, S. D. (1972). *J. Am. Chem. Soc.* **94**, 5924.

Kläui, W., and Neukomm, H. (1977). *Org. Magn. Reson.* **10**, 126.

Lang, R. W., and Hansen, H. J. (1979a). *Helv. Chim. Acta* **62**, 1025.

Lang, R. W., and Hansen, H. J. (1979b). *Helv. Chim. Acta* **62**, 1458.

Lau, K. S. Y., Wong, P. K., and Stille, J. K. (1976). *J. Am. Chem. Soc.* **98**, 5832.

Lefèvre, F., Burgemeister, T., and Mannschreck, A. (1977). *Tetrahedron Lett.*, p. 1125.

Lindoy, L. F., and Moody, W. E. (1977). *J. Am. Chem. Soc.* **99**, 5863.

Mannschreck, A., Jonas, V., and Kolb, B. (1973). *Agnew. Chem. Int. Ed. Engl.* **12**, 909.

Mannschreck, A., Jonas, V., Bödecker, H. O., Elke, H. L., and Kobrich, G. (1974). *Tetrahedron Lett.*, p. 2153.

McCreary, M. D., Lewis, D. W., Wernick, D. L., and Whitesides, G. M. (1974). *J. Am. Chem. Soc.* **96**, 1038.

Meyers, A. I., and Amos, R. A. (1980). *J. Am. Chem. Soc.* **102**, 870.

Meyers, A. I., and Yamamoto, Y. (1981). *J. Am. Chem. Soc.* **103**, 4278.

Meyers, A. I., Williams, D. R., Erickson, G. W., White, S., and Druelinger, M. (1981a). *J. Am. Chem. Soc.* **103**, 3081.

Meyers, A. I., Williams, D. R., White, S., and Erickson, G. W. (1981b). *J. Am. Chem. Soc.* **103**, 3088.

Midland, M. M., McDowell, D. C., Hatch, R. L., and Tramontano, A. (1980). *J. Am. Chem. Soc.* **102**, 867.

Mintas, M., Mannschreck, A., and Klasinc, L. (1981). *Tetrahedron* **37**, 867.

Mori, K. (1975). *Tetrahedron* **31**, 1381.

Mori, K. (1976). *Tetrahedron* **32**, 1979.

Morton, D. R., and Morge, R. A. (1978). *J. Org. Chem.* **43**, 2093.

Nagai, U., and Kobayashi, J. (1976). *Tetrahedron Lett.*, p. 2873.

Offermann, W., and Mannschreck, A. (1981). *Tetrahedron Lett.*, p. 3227.

Pirkle, W. II., and Sikkenga, D. L. (1975). *J. Org. Chem.* **40**, 3430.

Pluim, H., and Wynberg, H. (1980). *J. Org. Chem.* **45**, 2498.

Posner, G. H., Mallamo, J. P., and Miura, K. (1981). *J. Am. Chem. Soc.* **103**, 2886.

Raban, M., and Mislow, K. (1967). *Top. Stereochem.* **1**, 1.

Reger, D. L. (1975). *Inorg. Chem.* **14**, 660.

Reich, C. J., Sullivan, G. R., and Mosher, H. S. (1973). *Tetrahedron Lett.*, p. 1505.

Remy, D. C., Kettle, K. E., Hurst, C. A., Anderson, P. S., Arison, B. H., Engelhardt, E. L., Hirschmann, R., Clineschmidt, D. V., Lotti, V. J., Bunting, R. R., Ballantine, R. J., Papp, N. L., Flataker, L., Witoslawski, J. J., and Stone, C. H. (1977). *J. Med. Chem.* **20**, 1013.

Reuben, J. (1980). *J. Am. Chem. Soc.* **102**, 2232.

Sanders, J. K. M., and Williams, D. H. (1970). *J. Chem. Soc. Chem. Commun.*, p. 422.

Schollköpf, U., Groth, U., and Hartung, W. (1981). *Liebigs Ann. Chem.*, p. 2407.

Schurig, V. (1972). *Tetrahedron Lett.*, p. 3297.

Schurig, V. (1976). *Tetrahedron Lett.*, p. 1269.

Shaath, N. A., and Soine, T. O. (1975). *J. Org. Chem.* **40**, 1987.

Shapiro, B. L., and Johnston, M. D. (1972). *J. Am. Chem. Soc.* **94**, 8185.

Sievers, R. E. (1973). "Nuclear Magnetic Resonance Shift Reagents." Academic Press, New York.

Stackhouse, J., Cook, R. J., and Mislow, K. (1973). *J. Am. Chem. Soc.* **95**, 953.

Sullivan, G. R. (1976). *Top. Stereochem.* **10**, 287.

Sullivan, G. R., Ciaverella, D., and Mosher, H. S. (1974). *J. Org. Chem.* **39**, 2411.

Tangerman, A., and Zwanenburg, B. (1977). *Recl. Trav. Chim. Pays-Bas* **96**, 196.

Trost, B. M., and Strege, P. E. (1977). *J. Am. Chem. Soc.* **99**, 1650.

Trost, B. M., Shuey, C. D., DiNinno, F., Jr., and McElvain, S. S. (1979). *J. Am. Chem. Soc.* **101**, 1284.

Trost, B. M., O'Krongly, D., and Belletire, J. C. (1980). *J. Am. Chem. Soc.* **102**, 7595.

Valentine, D., Jr., Johnson, K. K., Priester, W., Sun, R. C. Toth, K., and Saucy, G. (1980). *J. Org. Chem.* **45**, 3698.

Von Ammon, R., and Fischer, R. D. (1972). *Agnew. Chem. Int. Ed. Engl.* **11**, 675.

Welti, D. H., Linder, M., and Ernst, R. R. (1978). *J. Am. Chem. Soc.* **100**, 403.

Wenzel, T. J., and Sievers, R. E. (1981). *Anal. Chem.* **53**, 393.

Wenzel, T. J., and Sievers, R. E. (1982). *J. Am. Chem. Soc.* **104**, 382.

Wenzel, T. J., Bettes, T. C., Sadlowski, J. E., and Sievers, R. E. (1980). *J. Am. Chem. Soc.* **102**, 5903.

Whitesides, G. M., and Lewis, D. W. (1970). *J. Am. Chem. Soc.* **92**, 6979.

Index